Math Mammoth
Grade 2-A Worktext

By Maria Miller

Copyright 2012 - 2025 Taina Maria Miller
ISBN 978-1480014763

2012 Edition

All rights reserved. No part of this book may be reproduced or transmitted in any form or by any means, electronic or mechanical, or by any information storage and retrieval system, without permission in writing from the author.

Copying permission: For having purchased this book, the copyright owner grants to the teacher-purchaser a limited permission to reproduce this material for use with his or her students. In other words, the teacher-purchaser MAY make copies of the pages, or an electronic copy of the PDF file, and provide them at no cost to the students he or she is actually teaching, but not to students of other teachers. This permission also extends to the spouse of the purchaser, for the purpose of providing copies for the children in the same family. Sharing the file with anyone else, whether via the Internet or other media, is strictly prohibited.

No permission is granted for resale of the material.

The copyright holder also grants permission to the purchaser to make electronic copies of the material for back-up purposes.

If you have other needs, such as licensing for a school or tutoring center, please contact the author at https://www.MathMammoth.com/contact

Contents

Foreword 6

Chapter 1: Some Old, Some New

Introduction 7
Some Review 10
The 100-Chart and More Review 12
Fact Families 15
Ordinal Numbers 17
Even and Odd Numbers 19
Doubling 21
One-Half 24
Adding with Whole Tens 26
Subtracting Whole Tens 29
Review Chapter 1 31

Chapter 2: Clock

Introduction 33
Review—Whole and Half Hours 37
The Minutes 38
The Minutes, Part 2 41
Past and Till in Five-Minute Intervals 43
How Many Hours Pass? 46
The Calendar: Weekdays and Months 48
The Calendar: Dates 51
Review Chapter 2 54

Chapter 3: Addition and Subtraction Facts Within 0-18

Introduction	55
Review: Completing the Next Whole Ten	59
Review: Going Over Ten	61
Adding with 9	63
Adding with 8	65
Adding with 7	67
Adding with 6	69
Review—Facts with 6, 7, and 8	71
Subtract to Ten	73
Difference and How Many More	75
Number Rainbows—11 and 12	78
Fact Families with 11	80
Fact Families with 12	81
Number Rainbows—13 and 14	83
Fact Families with 13 and 14	84
Fact Families with 15	87
Fact Families with 16	89
Fact Families with 17 and 18	91
Mixed Review Chapter 3	93
Review Chapter 3	95

Chapter 4: Regrouping in Addition

Introduction	98
Going Over to the Next Ten	101
Add with Two-Digit Numbers Ending in 9	104
Add a Two-Digit Number and a Single-Digit Number Mentally	106
Regrouping with Tens	108
Add in Columns Practice	111
Mental Addition of Two-Digit Numbers	114
Adding Three or Four Numbers Mentally	117
Adding Three or Four Numbers in Columns	119
Mixed Review Chapter 4	123
Review Chapter 4	125

Chapter 5: Geometry and Fractions

Introduction 127
Shapes Review 130
Surprises with Shapes 133
Rectangles and Squares 135
Making Shapes 138
Geometric Patterns 140
Solids 143
Printable Shapes 145
Some Fractions 153
Comparing Fractions 156
Mixed Review Chapter 5 158
Review Chapter 5 160

Foreword

Math Mammoth Grade 2 comprises a complete math curriculum for the second grade mathematics studies. The curriculum meets and exceeds the Common Core standards.

The main areas of study for second grade are:

1. Understanding of the base-ten system within 1000. This includes place value with three-digit numbers, skip-counting in fives, tens, and multiples of hundreds, tens, and ones (within 1000) (chapters 6 and 8);

2. Develop fluency with addition and subtraction, including solving word problems, regrouping in addition, and regrouping in subtraction (chapters 1, 3, 4, and 8);

3. Using standard units of measure (chapter 7);

4. Describing and analyzing shapes (chapter 5).

Additional topics we study are time, money, introduction to multiplication, and bar graphs and picture graphs.

This book, 2-A, covers reading the clock (chapter 2), the basic addition and subtraction facts within 18 (chapter 3), regrouping in addition (chapter 4), and geometry (chapter 5). The rest of the topics are covered in the 2-B student worktext.

Some important points to keep in mind when using the curriculum:

- These two books (parts A and B) are like a "framework", but you still have a lot of liberty in planning your child's studies. While addition and subtraction topics are best studied in the order they are presented, feel free to go through the sections on shapes, measurement, clock, and money in any order you like.

  This is especially advisable if your child is either "stuck" or is perhaps getting bored with some topic. Sometimes the concept the child was stuck on can become clear after a break from the topic.

- Math Mammoth is mastery-based, which means it concentrates on a few major topics at a time, in order to study them in depth. However, you can still use it in a *spiral* manner, if you prefer. Simply have your child study in 2-3 chapters simultaneously. This type of flexible use of the curriculum enables you to truly individualize the instruction for your child.

- Don't automatically assign all the exercises. Use your judgment, trying to assign just enough for your child's needs. You can use the skipped exercises later for review. For most children, I recommend to start out by assigning about half of the available exercises. Adjust as necessary.

- For review, the curriculum includes a worksheet maker (Internet access required), mixed review lessons, additional cumulative review lessons, and the word problems continually require usage of past concepts. Please see more information about review (and other topics) in the FAQ at
  https://www.mathmammoth.com/faq-lightblue.php

I heartily recommend that you view the full user guide for your grade level, available at
https://www.mathmammoth.com/userguides/

Lastly, you can find free videos matched to the curriculum at https://www.mathmammoth.com/videos/

> *I wish you success in teaching math!*
> Maria Miller, the author

Chapter 1: Some Old, Some New
Introduction

This chapter contains both some review and some new topics, with the aim of giving children a good start in second grade math.

In the first few lessons, we review adding and subtracting two-digit numbers, and skip-counting using the 100-chart, from first grade. Next, the lesson *Fact Families* reviews the connection between addition and subtraction, and introduces a new strategy for missing subtrahend problems (such as ___ − 5 = 4). In these problems, the child can *add* to find the missing total. This actually teaches them algebraic thinking.

Then we go on to the "new", starting with ordinal numbers, which are probably familiar from everyday language. Even and odd numbers are presented in the context of equal sharing: if you can share that many objects evenly (equally), then the number is even. Use manipulatives here if desired.

Then we study doubling and halving. Don't skip the word problems included in these lessons; they are important. Children need to learn to apply the concepts they have just learned. Also, if a child cannot solve word problems that involve doubling or halving, there is a chance they did not actually learn those concepts.

The last lessons have to do with adding and subtracting whole tens (multiples of ten) *mentally* (e.g. 51 + 30 or 72 − 40). Mental math is very important, because it builds number sense: the ability to manipulate numbers flexibly — to take them apart and put them together in various combinations. And number sense is very important: it actually predicts a student's success later on in algebra.

In this case, adding or subtracting multiples of ten is actually a concept rooted in place value. As long as the child understands place value (tens and ones), these types of problems are very easy. If your child has trouble, it is a sign they perhaps have not grasped place value with two-digit numbers.

Also, don't forget the free videos matched to the curriculum at https://www.mathmammoth.com/videos/.

Pacing Suggestion for Chapter 1

Please add one day to the pacing for the test if you use it. Note that the lessons in the chapter can take several days to finish. As a general guideline, second graders should finish 1.5-2 pages daily or 8-10 pages a week. Please also see the user guide at https://www.mathmammoth.com/userguides/ .

The Lessons in Chapter 1	page	span	suggested pacing	your pacing
Some Review	10	*2 pages*	2 days	
The 100-Chart and More Review	12	*3 pages*	2 days	
Fact Families	15	*2 pages*	1 day	
Ordinal Numbers	17	*2 pages*	1 day	
Even and Odd Numbers	19	*2 pages*	1 day	
Doubling	21	*3 pages*	2 days	
One-Half	24	*2 pages*	2 days	
Adding with Whole Tens	26	*3 pages*	2 days	
Subtracting Whole Tens	29	*2 pages*	1 day	
Review Chapter 1	31	*2 pages*	2 days	
Chapter 1 Test (optional)				
TOTALS		*23 pages*	16 days	

Games and Activities

Shuffle the Order

You need: Ten stuffed animals and a deck of number cards with numbers 1-10. Optionally: make a slide for the stuffed animals to slide down on.

Activity: Arrange the animals standing in a line, as if waiting for their turn to go on a slide. On your turn, draw two cards from the deck of number cards. The cards will act as ordinal numbers. The first card tells you which animal in line you will move, and the second card tells you to which position you move it to. For example, if you get 2 and 8, you will move the *second* animal to the *eighth* position in line.

After ten rounds, all the stuffed animals will go down the slide, *in order*.

Cover my Double

You need: One dice, two distinct kind of markers, such as pennies and dimes, or two kinds of beans. For a game board, draw a 4x4 grid with numbers 2, 4, 6, 8, 10, and 12 written multiple times.

Game Play: This is a game for two players. At your turn, throw the dice, and cover the double of what you get from the dice with one of your markers. Then it is the other player's turn. If the squares with your double are already covered, the turn passes to the other player. The winner is the first person who gets three of their markers in a row, or column, or diagonally.

4	2	10	8
8	4	8	12
6	12	10	6
2	6	4	8

Games and Activities at Math Mammoth Practice Zone

Hidden Picture Addition Game
Use a number range of 3 to 19, or some other, to practice addition.
https://www.mathmammoth.com/practice/mystery-picture

Hidden Picture Subtraction Game
Choose a number range of 2 to 18, for example, to practice subtraction in this fun game.
https://www.mathmammoth.com/practice/mystery-picture-subtraction

Two-Digit Mental Addition - Online Practice
Practice adding one two-digit number and one single-digit number without regrouping in this online quiz.
https://www.mathmammoth.com/practice/addition-subtraction-two-digit#opts=2p1dnr

Two-Digit Mental Subtraction - Online Practice
Practice subtracting a single-digit number from a two-digit number without regrouping in this online quiz.
https://www.mathmammoth.com/practice/addition-subtraction-two-digit#opts=2m1dnr

"7 Up" Card Game
You will see seven cards dealt face up. Simply choose any two cards that make 10 (or your chosen sum) to discard. When there are no cards that make that sum, click the deck to deal more cards. For this chapter, choose sums of 7, 8, 9, or 10.
https://www.mathmammoth.com/practice/seven-up

Skip-count in a 100-chart
Fill in numbers on a 100-chart in a specific skip-counting pattern. You can choose by which number to skip-count, and also the starting and ending numbers for the grid.
https://www.mathmammoth.com/practice/skip-count-hundred-chart

Fact Families

Choose which fact family or families to practice, and the program will give you addition and subtraction problems from those, including with missing numbers.

https://www.mathmammoth.com/practice/fact-families

Even and Odd

Sort numbers into even and odd by dragging each kind of number to its own "box" in this simple game.

https://www.mathmammoth.com/practice/even-or-odd

Fruity Math: Subtraction

Add a two-digit number and a multiple of ten (such as 57 − 20). Click the fruit with the correct answer and try to get as many points as you can within two minutes.

https://www.mathmammoth.com/practice/fruity-math#op=subtraction&duration=120&mode=manual&config=21,99x1__1,9x10&allow-neg=0

Fruity Math: Addition

Add a two-digit number and a multiple of ten (such as 26 + 30). Click the fruit with the correct answer and try to get as many points as you can within two minutes.

https://www.mathmammoth.com/practice/fruity-math#op=addition&duration=120&mode=manual&config=1,90x10__11,80x1&max-sum=100

Further Resources on the Internet

These resources match the topics in this chapter, and offer online practice, online games (occasionally, printable games), and interactive illustrations of math concepts. We heartily recommend you take a look. Many people love using these resources to supplement the bookwork, to illustrate a concept better, and for some fun. Enjoy!

https://links.mathmammoth.com/gr2ch1

Some Review

1. The box with a "T" means a TEN. The dots are ONES. Write the additions.

   a. [T T T] + [dots]   b. [T] + [T T / T T]   c. [T T / dots] + [T / dots]

   32 + 7 = 39   ___ + ___ = ___   ___ + ___ = ___

2. Add whole tens. To help, you can draw a ten-box or ten-boxes to the picture.

   a. 25 + 10 = ___   b. 14 + 10 = ___   c. 32 + 10 = ___

   25 + 20 = ___   14 + 20 = ___   32 + 20 = ___

   25 + 30 = ___   14 + 30 = ___   32 + 30 = ___

3. Subtract from 60 or from 30. One of the tens is shown with ten dots instead of a ten-box. Cover some of the dots to subtract.

   a. 60 − 3 = ___   b. 30 − 4 = ___

       60 − 8 = ___   30 − 6 = ___

       60 − 7 = ___   30 − 5 = ___

4. Add in columns. The two numbers to be added are shown with dots and ten-boxes.

   a.   b.

5. Subtract. In (a) and (b) you can cross out things in the picture to help you.

| a. [T][T] [T][T] :: :: 49 − 6 = ___ | b. [T][T] [T][T] :: :: 47 − 16 = ___ | c.   4 5
− 2 3 | d.   9 8
− 6 5 |

6. Add and subtract.

a.	b.	c.	d.
70 + 6 = ___	30 + 4 + 4 = ___	90 + ___ = 94	60 + ___ = 90
50 + 9 = ___	50 + 7 + 2 = ___	40 + ___ = 47	40 + ___ = 80
e.	**f.**	**g.**	**h.**
70 − 1 = ___	5 − 5 = ___	88 − 8 = ___	50 + ___ = 56
100 − 5 = ___	24 − 4 = ___	57 − 7 = ___	30 + ___ = 39

7. Solve the word problems.

a. Larry bought two boxes of crayons for $6 each, and some paper for $3.
What was the total cost?

b. Tom has seven marbles, and Leah has five. Leah gave Tom two of hers.
How many more marbles does Tom have now than Leah?

c. Phil has twenty shirts, and ten of them are white.
How many are not white?

d. A book costs $45. Can you buy it if you already have $22 and your grandma gives you another $20?

The 100-Chart and More Review

1. Skip-count by fives, starting at 5.
   Color these numbers light blue.

2. Skip-count by fives, starting at 6.
   Color these numbers yellow.

3. Skip-count by twos starting at 2, up to 30.
   Color these numbers pink.

4. Skip-count by twos backwards from 99 to 71. Color these numbers green.

5. Skip-count by fours starting at 4.
   Color these numbers yellow.
   It makes an interesting pattern!

6. Skip-count. First find by which number to skip-count, either by 2s, by 5s, or by 10s.

a. 40, 42, 44, ____, ____, 50, ____, ____, ____, ____, ____

b. ____, ____, ____, ____, 48, 58, 68, ____, 88, ____

c. ____, ____, ____, ____, 65, 63, 61, ____, ____, ____

d. ____, ____, ____, 70, 65, 60, ____, ____, ____, ____

7. Write the addition sentences. The box with a "T" is a ten. Under each problem, there is another, similar, addition problem for you to solve.

T T •• + ••	T T T •• + ••	T T T T •• + ••
a. ____ + ____ = ____	c. ____ + ____ = ____	e. ____ + ____ = ____
b. 34 + 3 = ____	d. 53 + 6 = ____	f. 32 + 5 = ____

8. Subtract by crossing some out. Under each problem, there is another problem that is similar.

T T T T T ••	T T T T ••	T T T T T ••
a. 59 − 6 = ____	c. 47 − 5 = ____	e. 60 − 3 = ____
b. 39 − 6 = ____	d. 67 − 5 = ____	f. 50 − 3 = ____

9. Add. The problems in each box are similar.

a.	b.	c.	d.
2 + 6 = ____	4 + 4 = ____	3 + 6 = ____	8 + 2 = ____
42 + 6 = ____	74 + 4 = ____	53 + 6 = ____	48 + 2 = ____
72 + 6 = ____	94 + 4 = ____	23 + 6 = ____	98 + 2 = ____

10. Subtract. The problems in each box are similar.

a.	b.	c.	d.
7 − 5 = ___	9 − 4 = ___	10 − 4 = ___	8 − 5 = ___
37 − 5 = ___	29 − 4 = ___	50 − 4 = ___	38 − 5 = ___
67 − 5 = ___	99 − 4 = ___	80 − 4 = ___	88 − 5 = ___

11. Add. In some of these problems you need to <u>make a new ten</u> with some of the little dots. You can also use a 100-bead abacus.

a. 17 + 8 = ___   b. 35 + 6 = ___   c. 24 + 16 = ___

d. 27 + 12 = ___   e. 19 + 24 = ___   f. 28 + 28 = ___

12. Find the number that goes into the shape.

a. 42 + 3 + ◯ = 50   b. 37 + ◯ + 1 = 40   c. 84 + ◯ + 4 = 90

13. Subtract the same number each time.

a. − 10

50	___
52	___
64	___
23	___

b. − 20

100	___
20	___
40	___
21	___

c. − 5

45	___
95	___
96	___
11	___

Fact Families

When two addition and two subtraction facts use the same numbers, it is called a *"fact family."*	$4 + 5 = \boxed{9}$ $5 + 4 = \boxed{9}$ $\boxed{9} - 5 = 4$ $\boxed{9} - 4 = 5$ Notice the TOTAL. The subtraction sentences <u>start</u> with the total.	$4 + 5 = 9$ $5 + 4 = 9$ $9 - 5 = 4$ $9 - 4 = 5$ Notice the PARTS. The two parts make up the total.
Sometimes in a subtraction problem, the *total* is asked: $\boxed{} - 8 = 20$ You know 20 and 8 are the "parts," and the total is missing. To find the total, just add the "parts": $20 + 8 = \underline{28}$		

1. Write two addition and two subtraction sentences—a fact family!

a.
___ + ___ = ___
___ + ___ = ___
___ − ___ = ___
___ − ___ = ___

b.
___ + ___ = ___
___ + ___ = ___
___ − ___ = ___
___ − ___ = ___

c.
___ + ___ = ___
___ + ___ = ___
___ − ___ = ___
___ − ___ = ___

2. Fill in the missing numbers. The four problems form a fact family.

a. $2 + \boxed{} = 8$
$\boxed{} + 2 = 8$
$8 - 2 = \boxed{}$
$8 - \boxed{} = 2$

b. ___ + ___ = 10
___ + ___ = 10
$10 - 7 = \boxed{}$
$10 - \boxed{} = 7$

c. ___ + ___ = ___
___ + ___ = ___
$9 - \boxed{} = 6$
___ − ___ = ___

3. Write a matching addition for the subtraction. There are two possibilities.

a. ___ + ___ = ___
8 − 2 = 6

b. ___ + ___ = ___
20 − 7 = 13

c. ___ + ___ = ___
60 − 20 = 40

When the first number is missing in a subtraction, it is the TOTAL that is missing.

You can find the TOTAL by adding the two numbers (those are the "parts").

☐ − 6 = 2

The total is missing. 6 and 2 are the "parts." So we add them. 2 + 6 = 8. The missing number is 8!

It is like "adding backwards":

$\overset{\text{Add.}}{\overset{+}{\leftarrow}}$
$\boxed{8} - 6 = 2$

$\overset{\text{Add.}}{\overset{+}{\leftarrow}}$
$\boxed{23} - 3 = 20$

---

4. The total is missing from the subtraction sentence. Solve.

a. ☐ − 5 = 4
b. ☐ − 7 = 2
c. ☐ − 7 = 10

5. Find the missing numbers.

a. ☐ − 2 = 4
☐ − 50 = 50
☐ − 8 = 20

b. ☐ − 7 = 80
60 + 4 = ☐
16 + ☐ = 20

c. 9 − ☐ = 5
77 + ☐ = 78
☐ − 9 = 60

Puzzle Corner

Find the missing numbers. This time adding backwards will NOT work!

a. 50 − ☐ = 10
33 − ☐ = 31

b. 100 − ☐ = 91
76 − ☐ = 72

c. 10 − ☐ − 2 = 1
9 − ☐ − 5 = 2

Ordinal Numbers

The numbers 1, 2, 3, 4, and so on are called cardinal numbers.

We also often use *ordinal* numbers. Ordinal numbers are used when talking about the *order* of things.

List of some ordinal numbers:

The *fourth* tree from the left is circled. It is also the *second* tree from the right.

M I S S I S S I P P I

The *seventh* letter of the word is S.

Ordinal Number	Name	Ordinal Number	Name
1st	first	9th	ninth
2nd	second	10th	tenth
3rd	third	11th	eleventh
4th	fourth	12th	twelfth
5th	fifth	13th	thirteenth
6th	sixth	14th	fourteenth
7th	seventh	15th	fifteenth
8th	eighth	16th	sixteenth

1. Circle.

   a. The second car from the left.

   b. The fifth car from the right.

   c. The seventh snowflake from the left.

   d. The fourth snowflake from the right.

   e. The ninth letter from the left.

   f. The twelfth letter from the right.

   E X T R A O R D I N A R Y

2. Color.

🌼🌼🌼🌼🌼 **a.** The third flower from the left	🌼🌼🌼🌼🌼 **b.** The first three flowers on the left
🌼🌼🌼🌼🌼🌼🌼 **c.** The fifth flower from the right.	🌼🌼🌼🌼🌼🌼🌼 **d.** The first five flowers on the right.

3. Find the letters, and find out what Jack's surprise gift was.

The second row from the top,
the first letter from the left. _____

The fourth row from the top,
the third letter from the left. _____

The first row from the top,
the fifth letter from the right. _____

The fifth row from the bottom,
the second letter from the right. _____

The 1st row from the bottom,
the 1st letter from the left. _____

The sixth row from the top,
the third letter from the right. _____

The 3rd row from the top,
the 2nd letter from the left. _____

The 1st row from the top,
the 2nd letter from the left. _____

E	S	L	A	B	G	P
B	H	E	N	I	V	S
W	N	K	P	T	L	F
J	D	A	U	D	W	M
Y	K	Z	N	Y	I	C
U	D	T	S	O	Q	R
O	T	H	A	V	E	L

4. **a.** Use letters from the given word to make a new word.

S U R P R I S I N G

___ ___ ___ ___ ___
10th 5th 6th 9th 1st

b. Put the letters in order to make a word. The first letter of your new word is "D."

N D Y R T C I A I O
7th 1st 10th 9th 4th 3rd 2nd 8th 5th 6th

D ___ ___ ___ ___ ___ ___ ___ ___ ___

Even and Odd Numbers

Can John and Jane share 4 balls evenly (so that both get as many balls)?

○ ○ ○ ○

Yes! Draw the balls for John and Jane.

John   Jane

Can John and Jane share 6 cars evenly? Try!

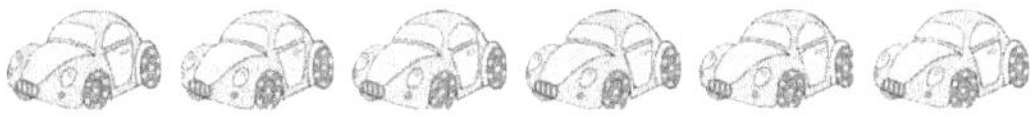

John   Jane

Can they share 5 carrots evenly? Try it!

Can they share 9 safety-pins evenly?

Four is an EVEN number because two people can share four things evenly.

Five is an ODD number because two people cannot share five things evenly.

1. Can two people share these things evenly? If yes, circle EVEN. If not, circle ODD.

10 marbles	7 marbles	3 stars
a. EVEN ODD	**b.** EVEN ODD	**c.** EVEN ODD
11 marbles	6 stars	4 marbles
d. EVEN ODD	**e.** EVEN ODD	**f.** EVEN ODD
9 stars	8 marbles	5 marbles
g. EVEN ODD	**h.** EVEN ODD	**i.** EVEN ODD

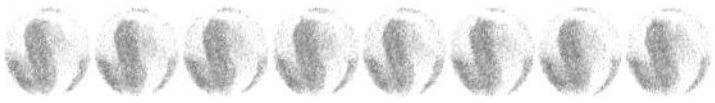

2. The chart shows how many cookies there are. Use rocks, beans, or other small items to make these amounts. Try to share them evenly with a friend. If you can share evenly, write "E" or "even" in the last column. If not, write "O" or "odd".

Cookies	Share evenly?	Even or odd?
11	NO	O
14		
15		

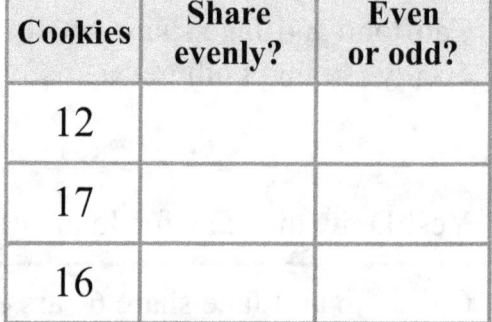

Cookies	Share evenly?	Even or odd?
12		
17		
16		

3. Color yellow all the EVEN numbers in the chart. Notice what pattern it makes! You can get help from your work in #1 and #2.

1	2	3	4	5	6	7	8	9	10
11	12	13	14	15	16	17	18	19	20

Now, color all the EVEN numbers in the rest of the 100-chart in the same pattern.

21	22	23	24	25	26	27	28	29	30
31	32	33	34	35	36	37	38	39	40
41	42	43	44	45	46	47	48	49	50
51	52	53	54	55	56	57	58	59	60
61	62	63	64	65	66	67	68	69	70
71	72	73	74	75	76	77	78	79	80
81	82	83	84	85	86	87	88	89	90
91	92	93	94	95	96	97	98	99	100

The numbers you didn't color are ODD numbers.

4. Look at the chart. Fill in.

Even numbers always end in (their last digit is) 2 , _____, _____, _____, or _____.

Odd numbers always end in (their last digit is) 1 , _____, _____, _____, or _____.

Doubling

> Doubling a number means adding it to itself. It is finding two times the number.
>
> **Examples:**
>
> Double 7 is 7 + 7 is <u>14</u>.   Double 20 is 20 + 20 is <u>40</u>.

1. Find the double of these numbers.

a. Double 4	**b.** Double 6	**c.** Double 8
___ + ___ = ___	___ + ___ = ___	___ + ___ = ___
d. Double 10	**e.** Double 30	**f.** Double 50
___ + ___ = ___	___ + ___ = ___	___ + ___ = ___

2. Find the double of these numbers by adding in the boxes.

    a. 22 + 22   **b.** Double 34   **c.** 13 + 13   **d.** Double 41

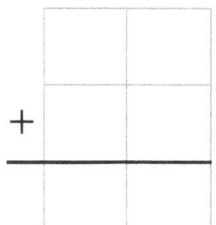 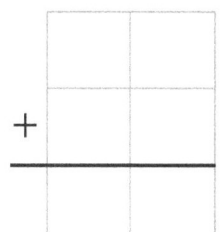 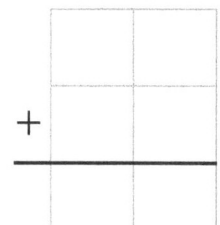 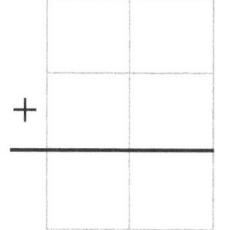

3. Make a doubles chart. Notice it has a pattern!

Double 1 = ___	6 + 6 = ___	11 + 11 = ___
Double 2 = ___	7 + 7 = ___	12 + 12 = ___
Double 3 = ___	8 + 8 = ___	13 + 13 = ___
Double 4 = ___	9 + 9 = ___	14 + 14 = ___
Double 5 = ___	10 + 10 = ___	15 + 15 = ___

> When you double a number, you always get an EVEN number as a result.
>
> Look at the ANSWER numbers in the doubles chart you just made.
> (You can color the answers yellow if you would like.)
>
> All of those numbers are EVEN numbers.
>
> If a number is even, you can share that many things evenly.

> **Example.** Double 13 is 13 + 13 = 26. This means that two children can share 26 toy cars EVENLY, and that each child gets 13 cars.

> **Example.** Two children need to clean 18 chairs. They divide the job equally (evenly). How many chairs does each child clean?
>
> Use the doubles chart. Since 9 + 9 = 18, each child will clean 9 chairs.

Solve the problems. You can use the doubles chart on the previous page to help you.

4. Mother tells two children to make 16 sandwiches.
   The children share the job equally.
   How many sandwiches will each child make?

5. You have 12 grapes and you share them evenly with your sister. How many do you get?

6. In a board game, you throw two dice and move that many spaces.
   Mary got double four. Andrea got double six.
   How many spaces did Mary move?

   How many spaces did Andrea move?

7. Cindy has 5 apples and Sandy has 3. They put them together and share them evenly. How many does each girl get?

8. Circle the even numbers: 13   20   19   8   15   16

Each number here is an even number, so it is a DOUBLE of some number. What number is it double of?

| 6 | 8 | 10 | 12 | 14 | 16 | 18 | 20 | 22 | 24 |

The first number on the list is 6. Six is double 3. We can write 6 = 3 + 3.
The last number on the list is 24. It is double 12. We can write 24 = 12 + 12.

9. Write each number as a double of some other number.

a. 8 = ___ + ___	b. 10 = ___ + ___	c. 4 = ___ + ___
d. 12 = ___ + ___	e. 14 = ___ + ___	f. 16 = ___ + ___

10. Write above each shaded number what number it is double of. Notice the pattern!

	5											
6	8	10	12	14	16	18	20	22	24	26	28	30

11. Mother and her friend need to make 20 dolls to sell.
    They share the job evenly.
    How many dolls will each woman make?

12. Two teachers divide 28 worksheets evenly.
    How many worksheets will each one get?

13. Mom found 7 cucumber slices in one container and 3 in another.
    You and your brother decide to share them equally.
    How many slices will you get?

14. (Challenge) A batch of brownies makes 16 brownies.
    Mom makes a double batch.
    How many brownies will she make?

One-Half

 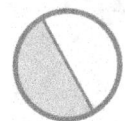

If you divide something into *two equal* parts, you have divided it into two halves. Each part is **half** of the whole.

Write one-half this way: $\frac{1}{2}$, or this way: 1/2.

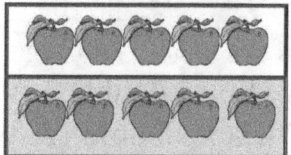

You can also find **half** of so many ***things***, if you have an <u>even</u> number of things.

5 + 5 = 10. So, half of ten apples is five apples.

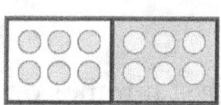

Twelve balls are divided into two equal parts. We can do that, because 12 is an even number.

6 + 6 = 12

$\frac{1}{2}$ of 12 is 6.

1. **a.** Color one half of each shape.   **b.** Color <u>two</u> halves of each shape.

2. Draw a line through these shapes and divide them into two halves. Color one half.

   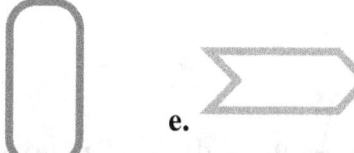

a. b. c. d. e.

3. Divide the things into two EQUAL groups. Write an addition. Find half of the total.

a. 10	b. 40	c. 24
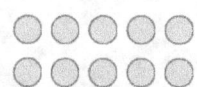	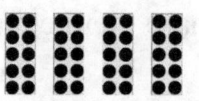	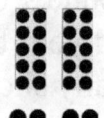
___ + ___ = ___	___ + ___ = ___	___ + ___ = ___
$\frac{1}{2}$ of 10 is ___.	$\frac{1}{2}$ of 40 is ___.	$\frac{1}{2}$ of 24 is ___.

Doubling and halving are the opposite things. $7 + 7 = 14$, so $\frac{1}{2}$ of 14 is 7.

4. Fill in the doubles chart. Then use it to find one-half of the given numbers.

6 + 6 = ____	11 + 11 = ____	$\frac{1}{2}$ of 16 is ____.
7 + 7 = ____	12 + 12 = ____	$\frac{1}{2}$ of 28 is ____.
8 + 8 = ____	13 + 13 = ____	$\frac{1}{2}$ of 26 is ____.
9 + 9 = ____	14 + 14 = ____	$\frac{1}{2}$ of 30 is ____.
10 + 10 = ____	15 + 15 = ____	$\frac{1}{2}$ of 22 is ____.

5. Divide the dots into two EQUAL groups. Find half of the total.

a. $\frac{1}{2}$ of 30 is ____.

b. $\frac{1}{2}$ of ____ is ____.

c. $\frac{1}{2}$ of ____ is ____.

6. Solve the problems. Then fill in another chart of doubles. *It has a pattern!* Find it!

   a. Jack and Joe split $60 evenly.
   How many dollars did each one get?

   b. Half of 100 students were sick.
   How many were not sick?

   c. Aunt Katie gave Missie half of $40.
   Missie spent $10 on a toy.
   How many dollars does Missie have now?

   d. The recipe called for 10 apples. That was exactly half of Mom's apples. How many apples did Mom have in the beginning?

   10 + 10 = ____
   15 + 15 = ____
   20 + 20 = ____
   25 + 25 = ____
   30 + 30 = ____
   35 + 35 = ____
   40 + 40 = ____

Adding with Whole Tens

1. The numbers are shown with ten-sticks and one-dots. Write the sums.

a. 54 + 10 = _____

b. _____ + 20 = _____

c. _____ + _____ = _____

d. _____ + _____ = _____

e. _____ + _____ = _____

f. _____ + _____ = _____

Adding whole tens and another 2-digit number Break down the other number into tens and ones. Add the tens. Then, add the ones.	50 + 26 50 + 20 + 6 70 + 6 = 76	39 + 40 30 + 9 + 40 70 + 9 = 79

2. Add. Break the second number into tens and ones first. Then add the tens.

a. 10 + 34 = _____ (10 + 30 + 4)	b. 10 + 28 = _____ (10 + _____ + _____)	c. 20 + 24 = _____ (20 + _____ + _____)
d. 30 + 21 = _____	e. 50 + 17 = _____	f. 40 + 33 = _____
g. 60 + 23 = _____	h. 30 + 37 = _____	i. 70 + 25 = _____

3. Add. Break the first number into tens and ones first. Then add the tens.

a. 45 + 20 = _____ (40 + 5 + 20)	b. 27 + 20 = _____ (___ + ___ + 20)	c. 45 + 40 = _____ (___ + ___ + 40)
d. 46 + 30 = _____	e. 16 + 50 = _____	f. 38 + 60 = _____
g. 20 + 77 = _____	h. 58 + 40 = _____	i. 40 + 39 = _____

4. Explain in your own words how you can mentally add 21 + 60.

5. Fill in the chart of doubles again, and notice its PATTERN.

5 + 5 = _____	30 + 30 = _____
10 + 10 = _____	35 + 35 = _____
15 + 15 = _____	40 + 40 = _____
20 + 20 = _____	45 + 45 = _____
25 + 25 = _____	50 + 50 = _____

6. Isabella got 30 books out of the library, and read half of them in two days. How many books does she have left to read?

7. Gwen and Mom went shopping. They bought shoes for $40, a top for $10, and a skirt for $20. Mom paid half of the cost and Gwen paid the rest. How much did Gwen pay?

8. Jacob had $61. Then he bought a toy for $30. How much money does he have left?

9. Fill in the missing numbers and find how many tens were added.

a. 12 + _____ = 22	b. 45 + _____ = 65	c. 23 + _____ = 63
12 + _____ = 52	45 + _____ = 55	23 + _____ = 53
12 + _____ = 42	45 + _____ = 75	23 + _____ = 93

10. Add 10, 20, 30, or 40. In the box below the number, write "E" if the number is even, and "O", if the number is odd. What do you notice?

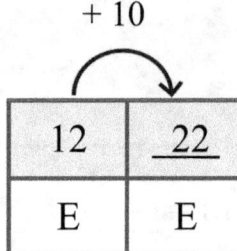 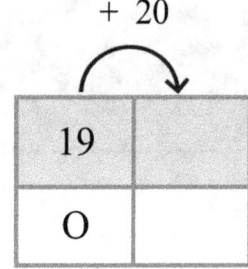 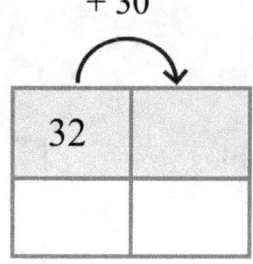 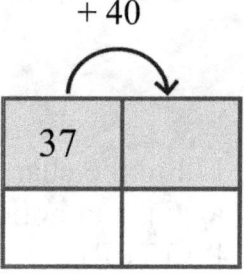

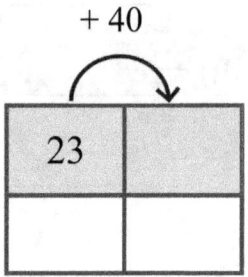 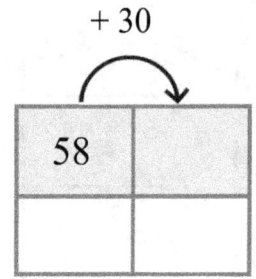 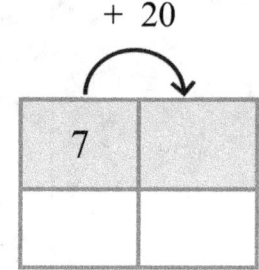 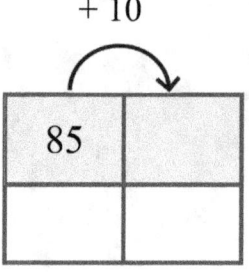

Puzzle Corner How many different solutions can you find for this puzzle? Find at least two. All numbers are whole tens.

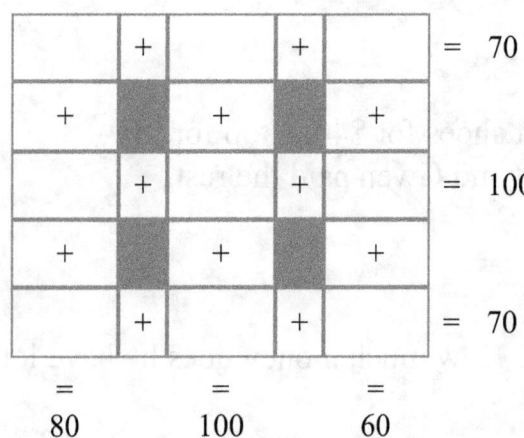 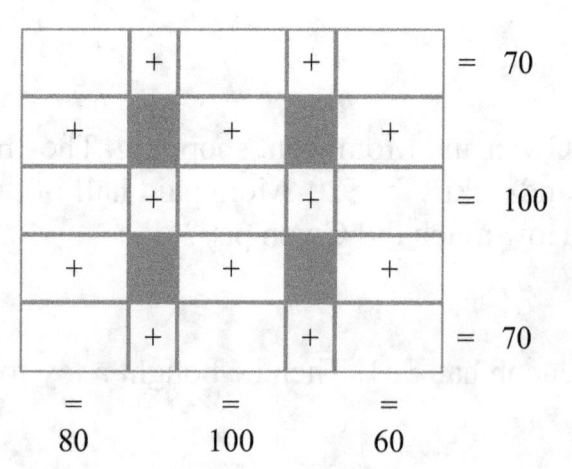

Subtracting Whole Tens

In the problem 47 − 20, think of the *tens*. The first number (47) has four tens. We take away two tens. So, there are TWO tens left.

The first number also has 7. That does not change.

Cross out two tens.

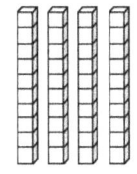

47 − 20 = _____

1. Cross out as many ten-pillars as the problem indicates. What is left?

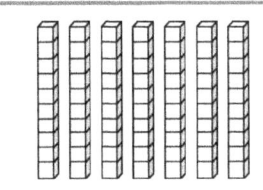

a. 70 − 50 = _____

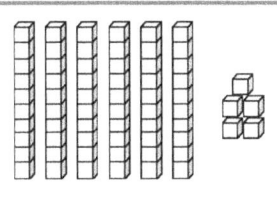

b. 65 − 30 = _____

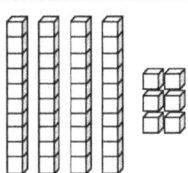

c. 46 − 20 = _____

Notice: The amount of ONES does not change in these subtractions.
You can just think of the TENS.

2. Count by tens backwards.

a. 76, 66, _____ , _____ , _____ , _____ , _____

b. _____ , _____ , 52, 42, _____ , _____ , _____

3. Subtract.

a.	b.	c.
23 − 10 = _____	48 − 20 = _____	56 − 10 = _____
23 − 20 = _____	48 − 30 = _____	56 − 30 = _____

d.	e.	f.
75 − 10 = _____	31 − 10 = _____	81 − 40 = _____
75 − 20 = _____	31 − 20 = _____	81 − 50 = _____

4. Find the pattern and continue it.

a. 88 − 10 = ___	b. 100 − 60 = ___	c. 34 − 10 = ___
88 − 20 = ___	90 − 50 = ___	44 − 20 = ___
88 − 30 = ___	80 − 40 = ___	54 − 30 = ___
88 − ___ = ___	___ − ___ = ___	___ − ___ = ___
88 − ___ = ___	___ − ___ = ___	___ − ___ = ___
88 − ___ = ___	___ − ___ = ___	___ − ___ = ___
88 − ___ = ___	___ − ___ = ___	___ − ___ = ___

5. Solve.

   a. Three suitcases weigh 30 kg, 18 kg, and 20 kg. How much is their total weight?

   b. Chairs cost $30 apiece. Can Dale buy three of them with $80?

   c. Henry received $50 for his birthday. If he buys three books that cost $10 each, how much will he have left?

Puzzle Corner Find numbers for the puzzles.

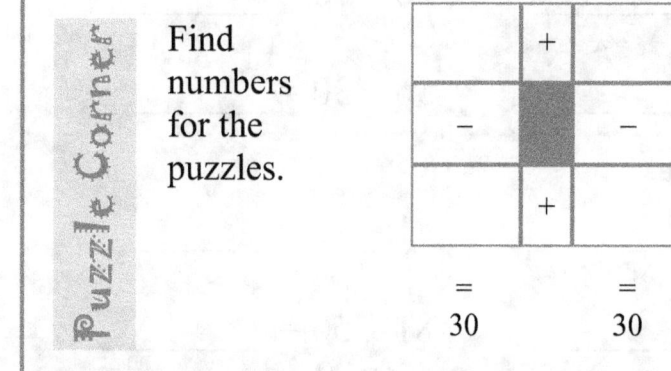
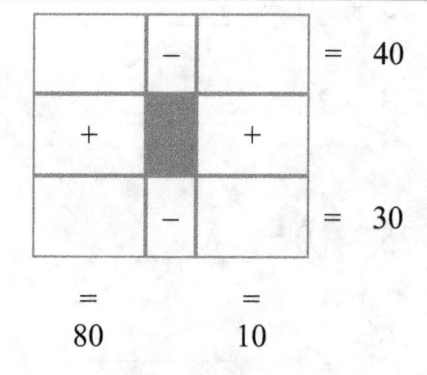

Review Chapter 1

1. Add. The problems in each box are similar.

a.	b.	c.	d.
51 + 7 = ___	46 + 3 = ___	72 + 5 = ___	35 + 5 = ___
81 + 7 = ___	96 + 3 = ___	32 + 5 = ___	95 + 5 = ___

2. Subtract. The problems in each box are similar.

a.	b.	c.	d.
49 − 5 = ___	29 − 3 = ___	60 − 7 = ___	38 − 4 = ___
89 − 5 = ___	69 − 3 = ___	80 − 7 = ___	78 − 4 = ___

3. **a.** How much would three shirts that cost $20 each cost together?

   b. Mike went to a yard sale and bought a desk for $32, a toy car for $1, a plant for $2, and some thread for $4. What was the total cost?

4. Add and subtract whole tens.

a.	b.	c.	d.
21 + 40 = ___	40 + 23 = ___	72 − 50 = ___	89 − 30 = ___
56 + 30 = ___	20 + 78 = ___	66 − 40 = ___	45 − 20 = ___

5. Use letters from the word W O N D E R F U L to make two new words.

___ ___ ___ ___   ___ ___ ___ ___
 1st   5th   9th   9th   4th   2nd   3rd   5th

6. Fill in the missing numbers. The four problems form a fact family.

a. 2 + ☐ = 10
    ☐ + 2 = 10
    10 − ___ = ☐
    10 − ☐ = ___

b. ___ + ___ = 9
    ___ + ___ = 9
    9 − 7 = ___
    9 − ___ = ___

c. ___ + ___ = ___
    ___ + ___ = ___
    8 − ___ = 5
    ___ − ___ = ___

7. The total is missing from the subtraction sentence. Solve.

a. ☐ − 8 = 8   b. ☐ − 5 = 4   c. ☐ − 30 = 30

8. Circle the even numbers.   72   31   59   60   8

9. Divide the dots into two EQUAL groups. Find half of the total.

a. $\frac{1}{2}$ of 50 is _____.

b. $\frac{1}{2}$ of 88 is _____.

c. $\frac{1}{2}$ of 46 is _____.

10. Two boys divided 18 toy cars evenly between them. How many did each boy get?

11. Mrs. Smith used half of her potatoes to make mashed potatoes. Now she has 13 potatoes left. How many did she have at first?

12. Mary has 13 colored pencils and Tina has twice as many. How many colored pencils do the girls have together?

Chapter 2: Clock
Introduction

The second chapter of *Math Mammoth Grade 2* deals with reading the clock to the five-minute intervals, and finding simple time intervals. I recommend having on hand an analog clock where the child can turn the hands on the clock.

First we practice telling time in the form of *hours:minutes* (such as 10:20), and then using the colloquial phrases "ten after," "quarter till," and so on.

Also studied are simple time intervals, or how much time passes. When practicing these topics, ask the child to move the minute (or hour) hand on an analog clock. The child can initially use a real clock for this, and later just imagine the movement of the clock hand(s) in his or her mind.

The chapter also has one lesson about the calendar. Of course, the calendar and the months are best learned simply in the context of everyday life, as the months pass. Hang a calendar on the wall and instruct your child to look at it every day, and to cross out days as they pass.

You can find several helpful videos that match these lessons at https://www.mathmammoth.com/videos/ .

If your child benefits from some slight spiraling, or just to keep things more interesting, feel free to mix the lessons in this chapter with the lessons in chapter 3, which is somewhat repetitive and can be more tedious to go through.

Pacing Suggestion for Chapter 2

Please add one day to the pacing for the test if you use it. Note that the specific lessons in the chapter can take several days to finish. They are not "daily lessons." As a general guideline, second graders should finish 1.5-2 pages daily or 8-10 pages a week. Please also see the user guide at https://www.mathmammoth.com/userguides/ .

The Lessons in Chapter 2	page	span	suggested pacing	your pacing
Review—Whole and Half Hours	37	*1 page*	1 day	
The Minutes	38	*3 pages*	2 days	
The Minutes, Part 2	41	*2 pages*	1 day	
Past and Till in Five-Minute Intervals	43	*3 pages*	2 days	
How Many Hours Pass?	46	*2 pages*	1 day	
The Calendar: Weekdays and Months	48	*3 pages*	2 days	
The Calendar: Dates	51	*3 pages*	2 days	
Review Chapter 2	54	*1 page*	1 day	
Chapter 2 Test (optional)				
TOTALS		*18 pages*	12 days	

Games and Activities

Tell the Time!

You need: An analog clock that allows you to turn the clock hands, or an app that allows you to do so.

In this activity, ask your child or student to turn the clock hands to a specific time (using the five-minute marks). Once they do so, then it is their turn to give you a time that you will set the clock to. You can use random times, and also important, specific times, such as, "We need to leave for the library at 2:45."

Find the Weekday!

You need: A wall calendar

In this simple activity, ask your child or student to find the weekday of a specific date on the calendar. Once they do so, then it is their turn to tell you what day of the week a certain date falls on. You can use random dates, and also important, specific dates, such as, "What day of the week is your birthday?"

How Many Months?

You need: A wall calendar

In this simple activity, ask your child or student to find how many months it is till someone's birthday, if right now it is a certain month. For example, let's say your birthday is in January. Ask, "If right now we're in June, how many months is it till my birthday?" Take turns, so that the child can ask you similar questions.

Earlier and Later

You need: A wall calendar or a calendar app

Ask your child or student to find the date one or two weeks after or before a certain date. For example, you could ask, "What date is it two weeks before September 6?" Take turns, so that the child will also ask you similar questions.

Months Match

This is a simple activity to practice matching the names of the months to their numbers.

You need: A set of 12 number cards with numbers from 1 to 12 on them. You can use cards from a standard deck if your child understands Jack as 11 and Queen as 12.

Shuffle the cards. Ask the child to turn the cards from the deck one by one, and at each card, say the name of the month that corresponds to that number. For example, if the child gets 7, they should say "July".

Once the child can go through all 12 cards without any mistakes, give them a small reward.

Games and Activities at Math Mammoth Practice Zone

Telling Time
Practice telling time on an analog clock with this interactive online exercise. Choose "To the nearest five minutes" for this grade level.
https://www.mathmammoth.com/practice/tell-time

Further Resources on the Internet

We have compiled a list of Internet resources that match the topics in this chapter, including pages that offer:

- **online practice** for concepts;
- online **games**, or occasionally, printable games;
- **animations** and interactive **illustrations** of math concepts;
- **articles** that teach a math concept.

We heartily recommend you take a look! Many of our customers love using these resources to supplement the bookwork. You can use these resources as you see fit for extra practice, to illustrate a concept better and even just for some fun. Enjoy!

https://l.mathmammoth.com/gr2ch2

Review—Whole and Half Hours

1. Write or say the time using the expressions *o'clock* or *half past*.

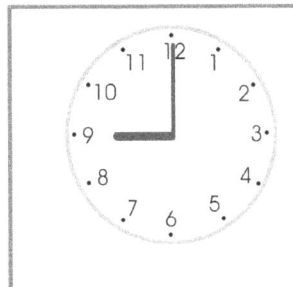   

a. _____ b. _____ c. _____ d. _____

2. Write the time in two ways: using the expressions *o'clock* or *half past*, and with numbers.

a. _____ o'clock   b. half past _____   c. half past _____   d. _____ o'clock

_____ : _____   _____ : _____   _____ : _____   _____ : _____

3. Write the time an hour later. Use numbers.

Now it is:	a. 6:00	b. 11:30	c. 3:00	d. 2:30	e. 9:30
An hour later, it is:					

4. Write the time a half-hour later. Use numbers.

Now it is:	a. 5:00	b. 7:30	c. 12:30	d. 10:00	e. 1:30
A half-hour later, it is:					

The Minutes

When the hour hand moves from one number to the next (from 1 to 2, or from 6 to 7, etc.), it takes one hour to do that.

In that same one hour of time, the *minute hand* travels **from 0 to 60 minutes**. So one hour is 60 minutes. A half-hour is 30 minutes.

When you read the minute hand, you use the green numbers (marked outside the clock face of the clock on the right). They go by fives, and are not normally marked on clocks. You need to know them. Just skip-count by fives!

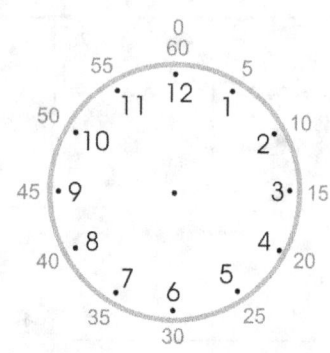

1 hour = 60 minutes.
1/2 hour = 30 minutes.

The hour hand is past 8.
The minute hand is at 15.
The time is 8:15.

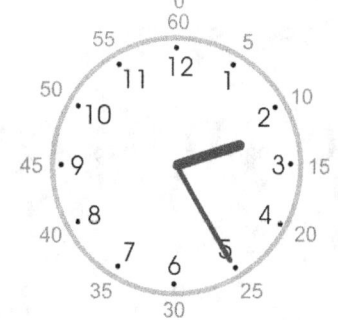

The hour hand is past 2.
The minute hand is at 25.
The time is 2:25.

The hour hand is past 11.
The minute hand is at 10.
The time is 11:10.

1. The arrow shows how much the minute hand travels. How many minutes pass?

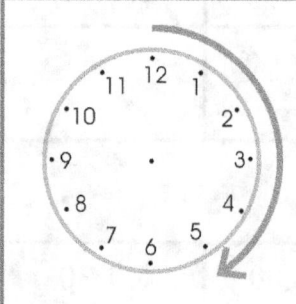

a. _____ minutes

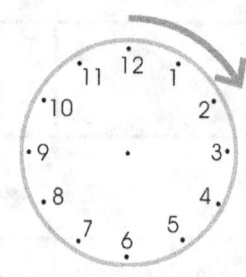

b. _____ minutes

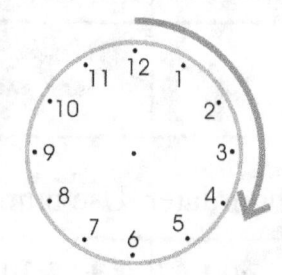

c. _____ minutes

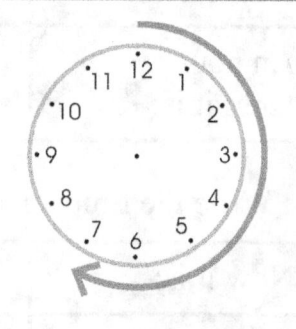

d. _____ minutes

2. Write the time. This special clock shows the numbers for hours *and* for minutes.

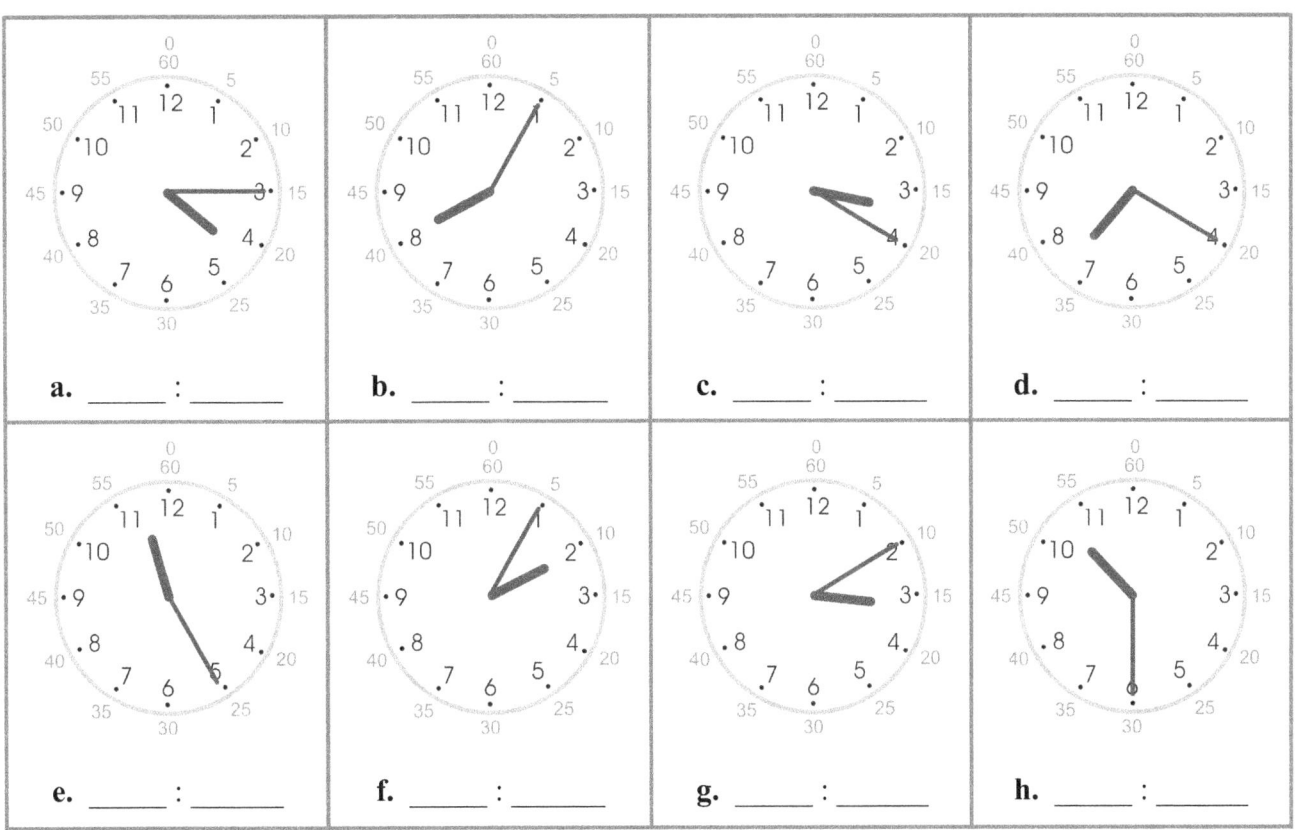

3. Write the time using the normal clock. Remember, the numbers for the minute hand are not shown, and they go by fives!

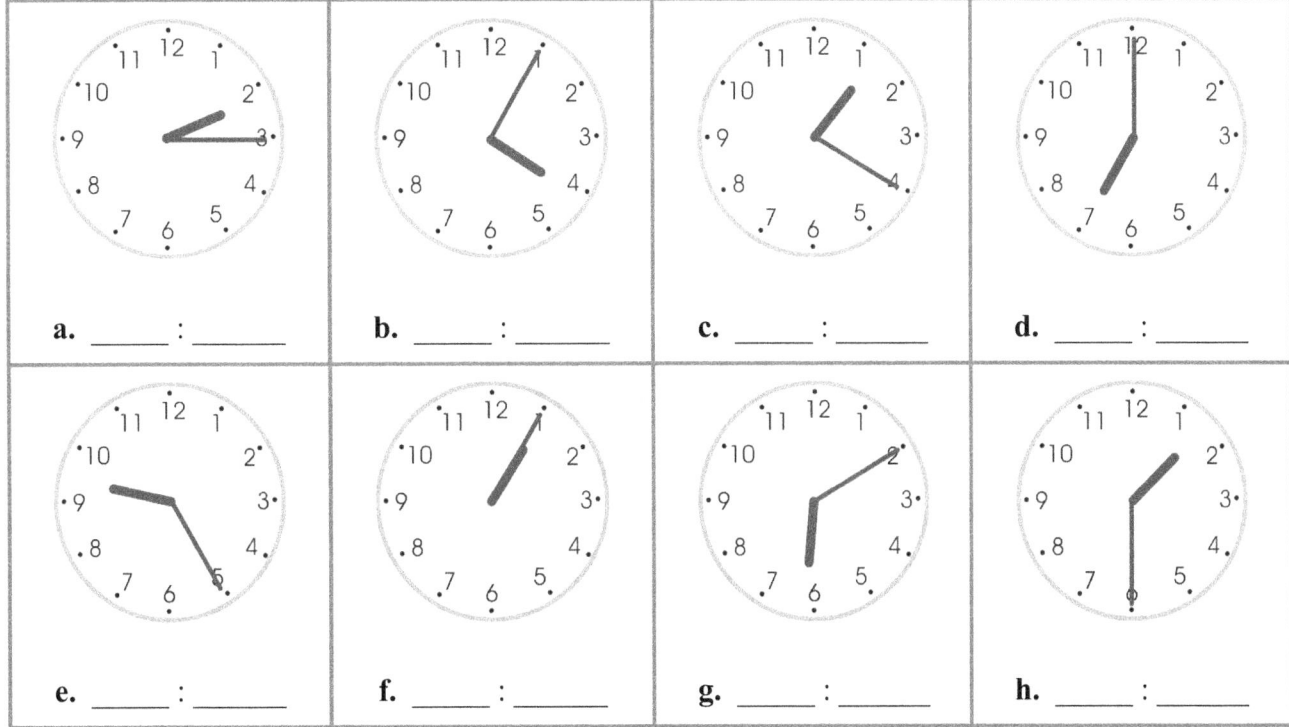

4. Find the clock that shows 11:25 and the clock that shows 11:05.

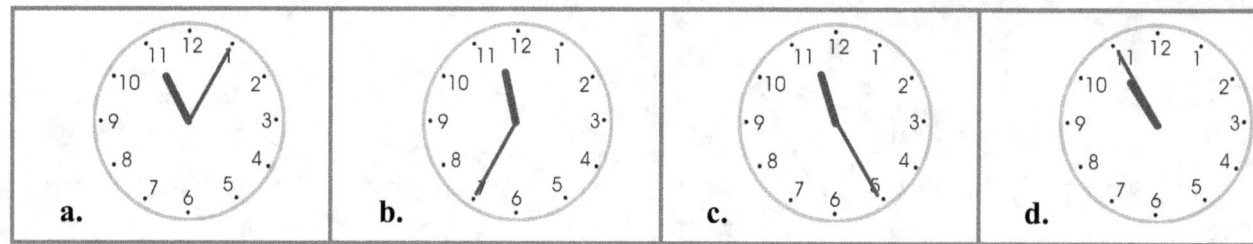

5. Write the time.

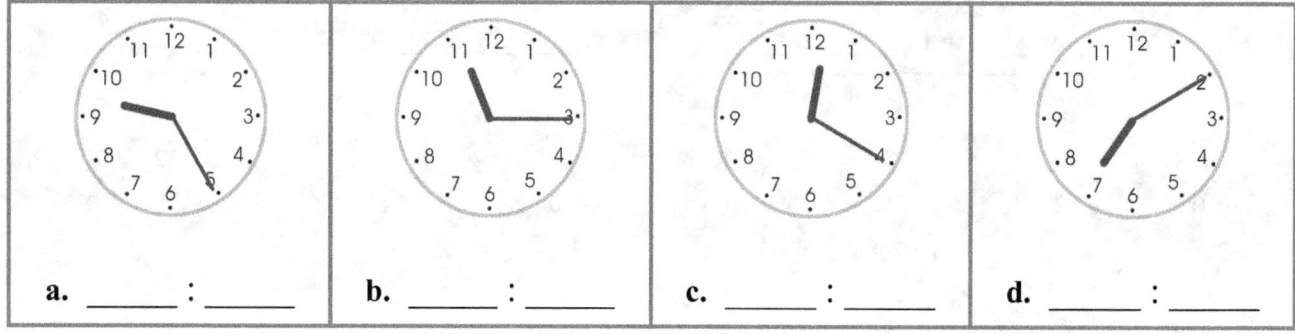

a. ____ : ____   b. ____ : ____   c. ____ : ____   d. ____ : ____

6. Write the time that the clock shows, and the time 5 minutes later. Imagine the minute hand moving one "step" further. You can use your practice clock.

	a. ____ : ____	b. ____ : ____	c. ____ : ____	d. ____ : ____
5 min. later →	____ : ____	____ : ____	____ : ____	____ : ____
	e. ____ : ____	f. ____ : ____	g. ____ : ____	h. ____ : ____
5 min. later →	____ : ____	____ : ____	____ : ____	____ : ____

The Minutes, Part 2

Notice! The hour hand *looks like* it is pointing to 2. But the minute hand has not yet reached 60 minutes, so it is *not yet* 2 o'clock!

We still say it is 1 hour (and some minutes).

The minute hand is at 45.
The time is **1**:45.

Another example. The hour hand *looks like* it is pointing to 10. But the minute hand has not yet reached 60 minutes, so it is *not yet* 10 o'clock.

We still say it is 9 hours (and some minutes).

The minute hand is at 55.
The time is **9**:55.

1. Choose the correct time.

a. Is it 1:50 or 2:50?

b. Is it 2:45 or 3:45?

c. Is it 6:55 or 7:55?

2. Draw the minute hand to match the given time. The hour hand is already drawn.

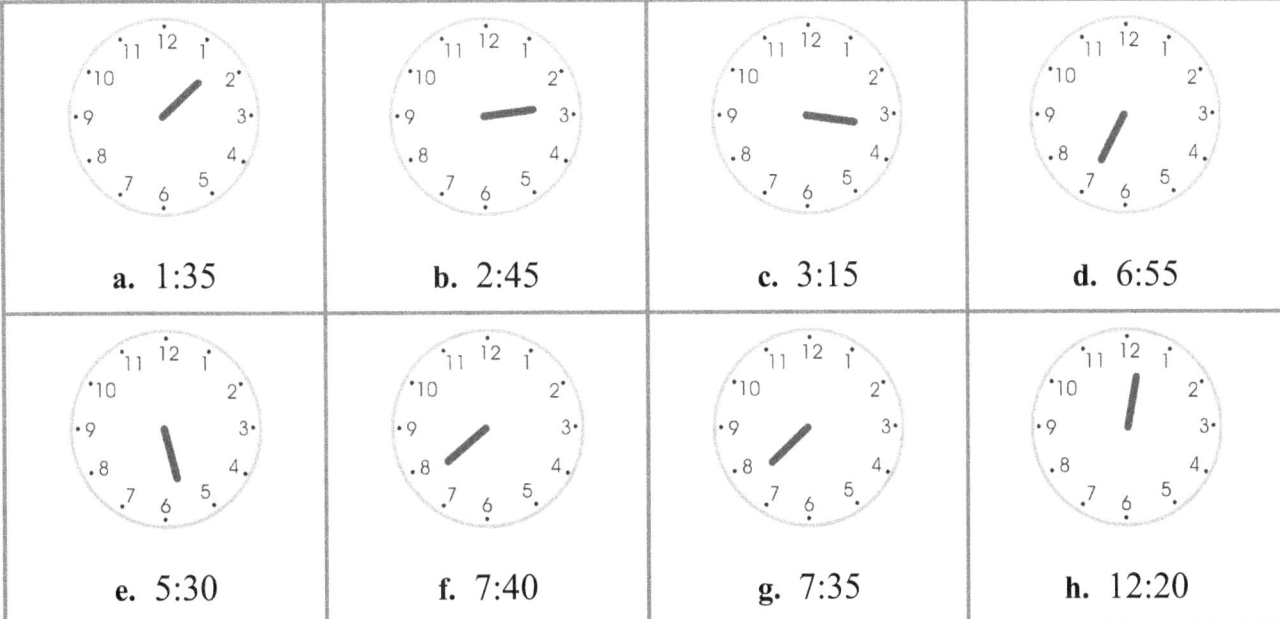

a. 1:35　　b. 2:45　　c. 3:15　　d. 6:55

e. 5:30　　f. 7:40　　g. 7:35　　h. 12:20

3. Write the time. Note: the hour hand *is close* to a number, but it has not reached it yet.

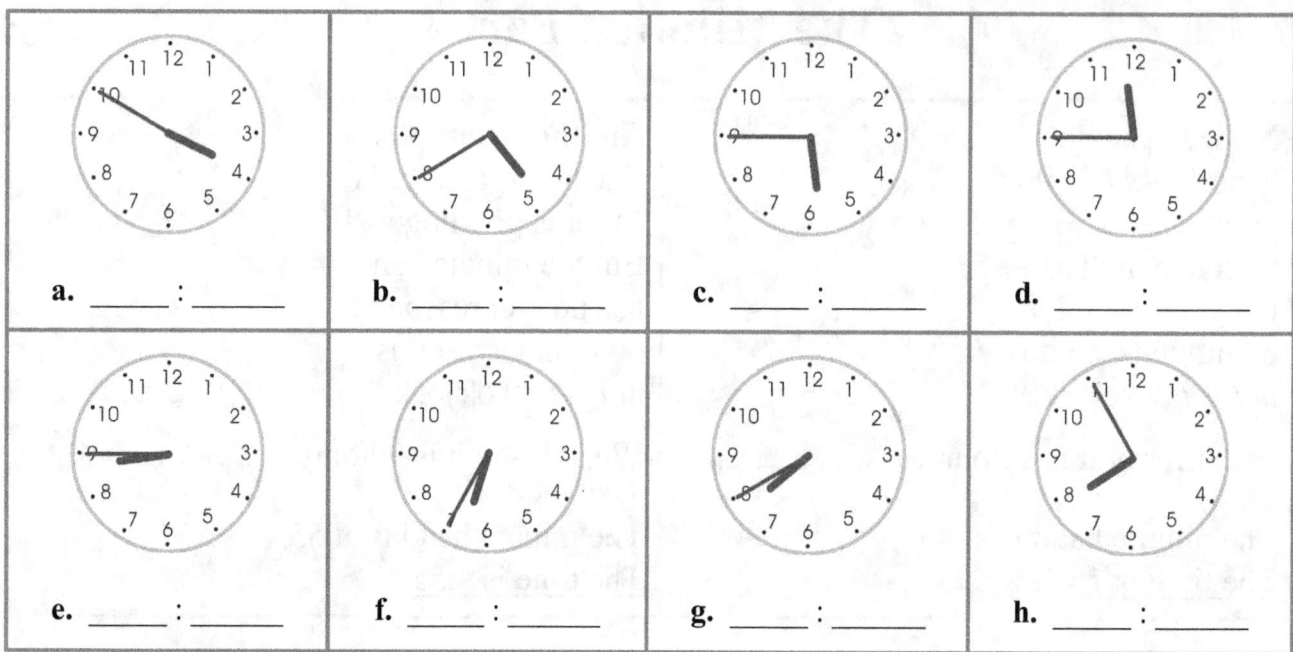

a. ___ : ___   b. ___ : ___   c. ___ : ___   d. ___ : ___

e. ___ : ___   f. ___ : ___   g. ___ : ___   h. ___ : ___

4. Write the time that the clock shows, and the time 5 minutes later.

	a.	b.	c.	d.
	___ : ___	___ : ___	___ : ___	___ : ___
5 min. later →	___ : ___	___ : ___	___ : ___	___ : ___

The children played 5-minute hide-and-seek, where they always took exactly 2 minutes for hiding and 3 for seeking. They used a watch to time it right.

After playing five rounds, Mom called them in to have an evening snack at 8:05. At what time did they start the game?

Past and Till in Five-Minute Intervals

We can also tell the time by saying how many minutes it is past the whole hour. Use the expression "so-many minutes past" **only** if the minutes are 30 or less.	 1:20 OR 20 past 1 "20 minutes past 1 o'clock".	 8:05 OR 5 past 8 "5 minutes past 8 o'clock".

1. How many minutes is it past the whole hour?

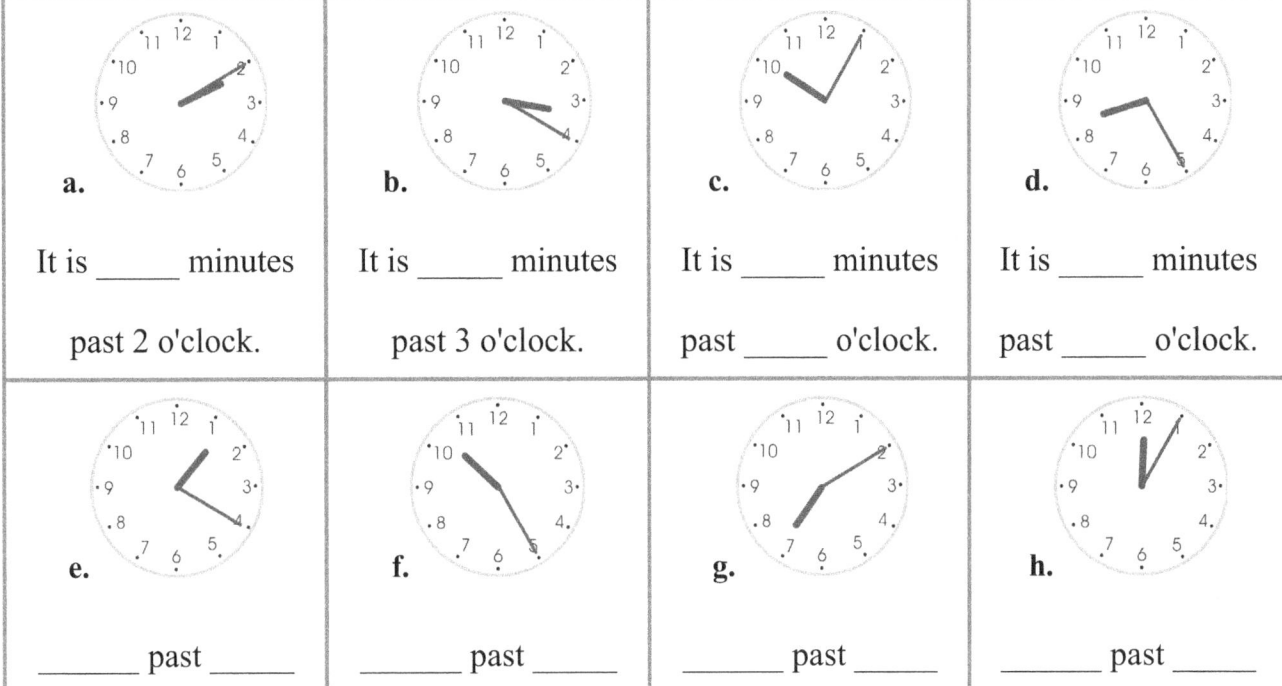

a. It is _____ minutes past 2 o'clock.

b. It is _____ minutes past 3 o'clock.

c. It is _____ minutes past _____ o'clock.

d. It is _____ minutes past _____ o'clock.

e. _____ past _____

f. _____ past _____

g. _____ past _____

h. _____ past _____

2. Write the time the clock shows and the time 5 minutes later. Use "past" or "half past".

	a. _____ past _____	b. _____ past _____	c. _____ past _____
5 min. later →			

We can also say how many minutes it is till the next whole hour.

Use this when the time is between half past some hour and the next whole hour (for example, between half past 6 and 7 o'clock).

To find the minutes till the next whole hour, count by fives — either from the next whole hour backwards to the minute hand, or vice versa.

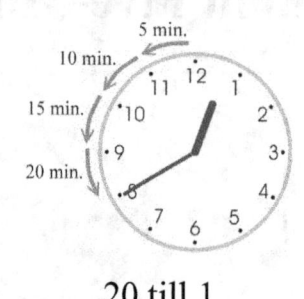

20 till 1

"20 minutes till 1 o'clock."

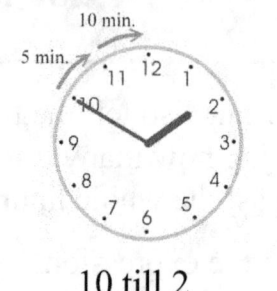

10 till 2

"10 minutes till 2 o'clock."

Notice that in this wording we use the *next* whole hour—the hour that is still yet to come.

The time 6:35 looks quite different from 25 till 7, but they mean the same. The time 6:35 shows the hour that the hour hand has <u>passed</u> (six), whereas 25 till 7 shows the hour that the hour hand is <u>coming to</u> (seven), or the *next* hour.

6:35   OR   25 till 7

3:55   OR   5 till 4

3. Show how many minutes it is till the next whole hour.

a. It is ____ minutes till 7 o'clock.

b. It is ____ minutes till ____ o'clock.

c. It is ____ minutes till ____ o'clock.

d. It is ____ minutes till ____ o'clock.

e. ____ till ____

f. ____ till ____

g. ____ till ____

h. ____ till ____

i. ____ till ____

j. ____ till ____

k. ____ till ____

l. ____ till ____

4. Write the time using the wordings "past" or "till", and using numbers.

a.	b.	c.
_____	_____	_____
_____ : _____	_____ : _____	_____ : _____
d.	e.	f.
_____	_____	_____
_____ : _____	_____ : _____	_____ : _____

5. Write the time using the **hours:minutes** way. Use your practice clock to help.

a. 10 past 8	b. 15 till 7	c. 25 past 12	d. half-past 7
_____ : _____	_____ : _____	_____ : _____	_____ : _____
e. 9 o'clock	f. 20 till 6	g. 5 till 11	h. 25 till 4
_____ : _____	_____ : _____	_____ : _____	_____ : _____

6. Write the time using the expressions "past," "till," or "half past."

a. 6:45 _____

b. 9:30 _____

c. 12:10 _____

d. 4:55 _____

e. 8:35 _____

f. 1:40 _____

How Many Hours Pass?

The chart below shows the whole hours in one 24-hour period = one night + one day.

12 – 1 – 2 – 3 – 4 – 5 – 6 – 7 – 8 – 9 – 10 – 11 – 12 – 1 – 2 – 3 – 4 – 5 – 6 – 7 – 8 – 9 – 10 – 11 – 12
Midnight   AM   Noon   PM   Midnight

From midnight to noon we call the hours "AM". This comes from *Ante Meridiem* (Latin), and means *before noon*. From noon to midnight we call the hours "PM", which comes from *Post Meridiem* (Latin), and means *after noon*.

How many hours is it from 6 AM to 11 AM?

You could use the chart, and count. But since both hours are AM, you can use subtraction to find the difference: 11 − 6 = 5 hours.

How many hours is it from 3 AM to 3 PM?

Now you cannot use subtraction because the answer clearly is not zero hours. Since the number is the same (3), it means the hour hand travels through the entire clock face, starting at 3 and ending at 3. The difference is 12 hours.

How many hours is it from 8 AM to 3 PM?

One of the times is AM, and the other is PM, so you cannot subtract them. Instead, do it in two parts:

1) How many hours from 8 AM till noon? It is four hours.
2) How many hours from noon till 3 PM? It is three hours.

All totaled, there are 7 hours from 8 AM to 3 PM.

1. How many hours is it?

from	5 AM	7 AM	9 AM	11 AM	10 AM
to	12 noon	1 PM	4 PM	11 PM	7 PM
hours					

2. How long is the school day, if it starts and ends at given times?

Start:	8 AM	8 AM	9 AM	10 AM	8 AM
End:	12 noon	1 PM	3 PM	3 PM	2 PM
hours:					

3. How many hours is it till midnight?

from	4 PM	7 PM	12 noon	9 AM	7 AM
to	12 midnight	12 midnight	12 midnight	12 midnight	12 midnight
hours					

4. How many hours does Matthew sleep if he goes to bed and gets up at given times?

Go to bed	9 PM	8 PM	9 PM	11 PM	12 midnight
Get up	6 AM	7 AM	5 AM	9 AM	9 AM
Sleep hours					

5. **a.** How many hours is your school day usually?

   b. How many hours do you usually sleep?

6. **a.** Dad's workday starts at 8:00 in the morning, and ends at 5 PM.
   How many hours is Dad at work?

   b. Mary's school day starts at 9 AM and ends at 2 PM. How long is her school day?

   c. The airplane took off at 10 AM and landed at 1 PM. Then it took off again at 2 PM and landed at 6 PM. How many hours was the airplane in the air?

7. **a.** How many hours are there in one day-night period?

   b. How many hours are there in two day-night periods?

8. **a.** The turkey needs to cook three hours in the oven to be ready at 7 PM.
   When should it be put into the oven?

   b. It takes two hours to mow the lawn. Jim wants to be done at 1 PM.
   When should he start mowing?

   c. Mom needs seven hours of sleep tonight. She wants to wake up at 6 AM.
   When should she go to bed?

The Calendar: Weekdays and Months

Calendar		
January Su Mo Tu We Th Fr Sa 1 2 3 4 5 6 7 8 9 10 11 12 13 14 15 16 17 18 19 20 21 22 23 24 25 26 27 28 29 30 31	**February** Su Mo Tu We Th Fr Sa 1 2 3 4 5 6 7 8 9 10 11 12 13 14 15 16 17 18 19 20 21 22 23 24 25 26 27 28	**March** Su Mo Tu We Th Fr Sa 1 2 3 4 5 6 7 8 9 10 11 12 13 14 15 16 17 18 19 20 21 22 23 24 25 26 27 28 29 30 31

April
 	
Su Mo Tu We Th Fr Sa
          1 2 3 4
 5 6 7 8 9 10 11
12 13 14 15 16 17 18
19 20 21 22 23 24 25
26 27 28 29 30
	May
	
	Su Mo Tu We Th Fr Sa
	                   1 2
	 3 4 5 6 7 8 9
	10 11 12 13 14 15 16
	17 18 19 20 21 22 23
	24 25 26 27 28 29 30
	31
		June
		
		Su Mo Tu We Th Fr Sa
		      1 2 3 4 5 6
		 7 8 9 10 11 12 13
		14 15 16 17 18 19 20
		21 22 23 24 25 26 27
		28 29 30

July
 	
Su Mo Tu We Th Fr Sa
          1 2 3 4
 5 6 7 8 9 10 11
12 13 14 15 16 17 18
19 20 21 22 23 24 25
26 27 28 29 30 31
	August
	
	Su Mo Tu We Th Fr Sa
	                       1
	 2 3 4 5 6 7 8
	 9 10 11 12 13 14 15
	16 17 18 19 20 21 22
	23 24 25 26 27 28 29
	30 31
		September
		
		Su Mo Tu We Th Fr Sa
		        1 2 3 4 5
		 6 7 8 9 10 11 12
		13 14 15 16 17 18 19
		20 21 22 23 24 25 26
		27 28 29 30

October
 	
Su Mo Tu We Th Fr Sa
                1 2 3
 4 5 6 7 8 9 10
11 12 13 14 15 16 17
18 19 20 21 22 23 24
25 26 27 28 29 30 31
	November
	
	Su Mo Tu We Th Fr Sa
	 1 2 3 4 5 6 7
	 8 9 10 11 12 13 14
	15 16 17 18 19 20 21
	22 23 24 25 26 27 28
	29 30
		December
		
		Su Mo Tu We Th Fr Sa
		            1 2 3 4 5
		 6 7 8 9 10 11 12
		13 14 15 16 17 18 19
		20 21 22 23 24 25 26
		27 28 29 30 31

1. You see "Su Mo Tu We Th Fr Sa" in the calendar. What does it mean?
   Ask your teacher if you don't know!
   _____

   They are called the "days of the week." Learn to say them from memory.

2. Fill in the weekday before and after the given day. Try NOT to look at the calendar!

	Tuesday	
	Friday	
	Sunday	

3. What day of the week is it today? _____

4. Let's say it is Friday. What day of the week will it be...

   a. in 1 day? _____ **b.** in 7 days? _____

   c. in 5 days? _____ **d.** in 3 days? _____

5. Look at the calendar on the previous page. Tell what day of the week is...

   your birthday   _____

   January 1
    (New Year's Day)   _____

   May 10 (Mother's Day)   _____

   September 7 (Labor Day)   _____

   _____   _____

   _____   _____

   _____   _____

   (Fill in some other dates of your choosing for the last three lines.)

6. **a.** Circle all the months that have 31 days.

> January February March April May June July August September October November December

b. Circle all the months that have 30 days.

> January February March April May June July August September October November December

c. Which month didn't get circled either time? _____

It normally has 28 days, but every four years (each LEAP year) it has 29 days.

7. Fill in the month before and after the given month. Try NOT to look at the calendar!

	March	
	August	
	November	

> In April, Mrs. Warwick sent a package to her friend in China. It took a long time to arrive, and arrived in August. How many months did the package take?
>
> Count up the months till August, but don't start at April; start at the month just after April (which is May). *May, June, July, August.* You counted up four months. The package took four months to arrive.

8. Let's say it is JUNE now. Children figure out how long it is until their birthday. Count up the months, and fill in the blanks.

   a. Anna's birthday is in September. It is still _____ months till Anna's birthday.

   b. Kyle's birthday is in August. It is only _____ months till Kyle's birthday.

   c. May's birthday is in December. It is _____ months till May's birthday.

9. How about you? In what month is your birthday? _____

   How many months is it till your birthday this year? _____ months

   Or, if you already had it, how many months ago was it? _____ months ago

The Calendar: Dates

Calendar		

January
Su Mo Tu We Th Fr Sa
                    1 2
3 4 5 6 7 8 9
10 11 12 13 14 15 16
17 18 19 20 21 22 23
24 25 26 27 28 29 30
31
	February
	Su Mo Tu We Th Fr Sa
	    1 2 3 4 5 6
	7 8 9 10 11 12 13
	14 15 16 17 18 19 20
	21 22 23 24 25 26 27
	28 29
		March
		Su Mo Tu We Th Fr Sa
		       1 2 3 4 5
		6 7 8 9 10 11 12
		13 14 15 16 17 18 19
		20 21 22 23 24 25 26
		27 28 29 30 31

April
Su Mo Tu We Th Fr Sa
                    1 2
3 4 5 6 7 8 9
10 11 12 13 14 15 16
17 18 19 20 21 22 23
24 25 26 27 28 29 30
	May
	Su Mo Tu We Th Fr Sa
	1 2 3 4 5 6 7
	8 9 10 11 12 13 14
	15 16 17 18 19 20 21
	22 23 24 25 26 27 28
	29 30 31
		June
		Su Mo Tu We Th Fr Sa
		          1 2 3 4
		5 6 7 8 9 10 11
		12 13 14 15 16 17 18
		19 20 21 22 23 24 25
		26 27 28 29 30

July
Su Mo Tu We Th Fr Sa
                    1 2
3 4 5 6 7 8 9
10 11 12 13 14 15 16
17 18 19 20 21 22 23
24 25 26 27 28 29 30
31
	August
	Su Mo Tu We Th Fr Sa
	    1 2 3 4 5 6
	7 8 9 10 11 12 13
	14 15 16 17 18 19 20
	21 22 23 24 25 26 27
	28 29 30 31
		September
		Su Mo Tu We Th Fr Sa
		              1 2 3
		4 5 6 7 8 9 10
		11 12 13 14 15 16 17
		18 19 20 21 22 23 24
		25 26 27 28 29 30

October
Su Mo Tu We Th Fr Sa
                       1
2 3 4 5 6 7 8
9 10 11 12 13 14 15
16 17 18 19 20 21 22
23 24 25 26 27 28 29
30 31
	November
	Su Mo Tu We Th Fr Sa
	       1 2 3 4 5
	6 7 8 9 10 11 12
	13 14 15 16 17 18 19
	20 21 22 23 24 25 26
	27 28 29 30
		December
		Su Mo Tu We Th Fr Sa
		              1 2 3
		4 5 6 7 8 9 10
		11 12 13 14 15 16 17
		18 19 20 21 22 23 24
		25 26 27 28 29 30 31

1. Mary goes swimming every Thursday. Look at the calendar, and write the **May** dates when Mary goes swimming. Use the form (*month*) (*day*), such as May 5.

_____ _____   _____ _____

_____ _____   _____ _____

2. Circle these public and school holidays on the calendar on the previous page.

January 1	New Year's Day
January 18	Martin Luther King Day
February 15	President's Day
March 21-25	Spring break
May 30	Memorial Day
July 4	Independence Day

September 5	Labor Day
October 10	Columbus Day (observed)
November 11	Veteran's Day
November 24	Thanksgiving
Dec. 22- Jan. 2	Winter break

3. For this exercise, you need a current calendar. Look at this year's calendar and write the dates below in the form (*month*) (*day*) (*year*), for example as June 15, 2016.

	month	day	year
a. today's date			
b. tomorrow's date			
c. your birthday this year			
d. Christmas day, this year			
e. the first Monday of June			
f. the last Friday of August			

4. Cindy sent a letter to her friend on October 25th. The letter took two days to reach her friend. What date did her friend get it?

5. Julie got glasses in June. The eye doctor told her to come back in four months. Count four months, starting your count at the month after June. In what month will Julie go back to the eye doctor?

6. The soccer team played their last game in late November, and then they took a 2-month break. In what month did they start playing again?

On the calendar, October 14 is highlighted. The date *one week ago* is just above that: it is October 7 (underlined).

The date *one week later* than October 14 is just below that: it is October 21 (underlined).

What would the date be *two weeks later* than October 14?

October
Su	Mo	Tu	We	Th	Fr	Sa
						1
2	3	4	5	6	**7**	8
9	10	11	12	13	**14**	15
16	17	18	19	20	**21**	22
23	24	25	26	27	28	29
30	31					

What date is 1 week later than October 29?

October 29 is a Saturday, so the date one week later is also a Saturday. It will be the first Saturday of November. That is November 5.

What would be the date *two weeks later* than October 29?

October
Su	Mo	Tu	We	Th	Fr	Sa
						1
2	3	4	5	6	7	8
9	10	11	12	13	14	15
16	17	18	19	20	21	22
23	24	25	26	27	28	**29**
30	31					

November
Su	Mo	Tu	We	Th	Fr	Sa
		1	2	3	4	**5**
6	7	8	9	10	11	12
13	14	15	16	17	18	19
20	21	22	23	24	25	26
27	28	29	30			

7. Fill in the missing dates in the table. Use the calendar to help!

Date 1 week ago	Date now	Date 1 week later
	July 14	
	December 8	
	January 26	

Date 2 weeks ago	Date now	Date 2 weeks later
	August 8	
	October 18	
	February 23	

8. The painting class meets every two weeks. Their first meeting of the year is January 12. What are the dates for the next two meetings?

Review Chapter 2

1. Write the time with *hours:minutes*, and using "past", "till", "half past" or "o'clock".

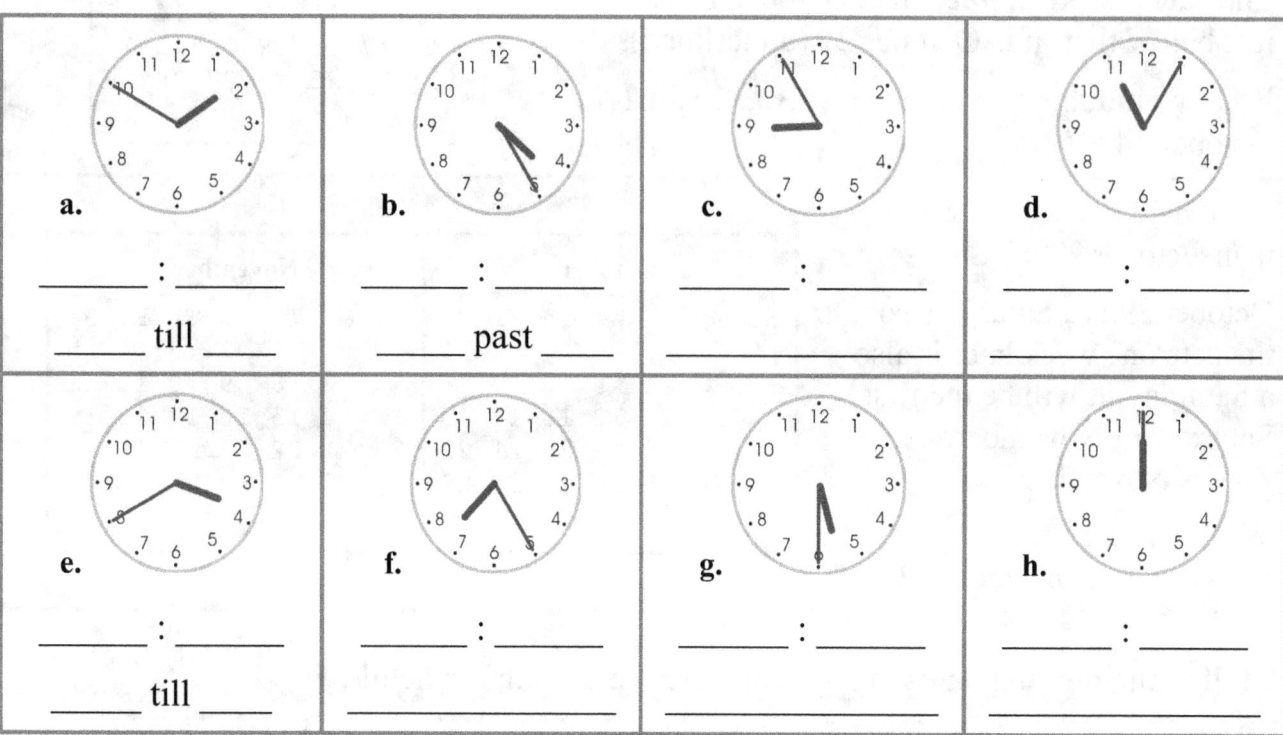

2. Write the later time.

Time now	2:30	6:55
5 min. later	____:____	____:____

Time now	9:05	5:40
10 min. later	____:____	____:____

3. Father starts his work at 9 AM, and leaves to go back home at 5 PM. How many hours is his work day?

4. Little Emily goes to kindergarten at 8 AM and stays there four hours. At what time does her kindergarten class end?

5. Jonah goes to the chess club every Thursday. He went today, March 17. What is the date when he goes next time?

6. Emily got the August issue of a magazine in the mail. The next magazine comes in three months. What month will that be?

Chapter 3: Addition and Subtraction Facts Within 0-18
Introduction

The third chapter of *Math Mammoth Grade 2* provides lots of practice for learning and memorizing the basic addition and subtraction facts of single-digit numbers with an answer between 10 and 18.

The goal is to memorize the facts, or at least become so fluent with them that an outsider cannot tell if the child remembers the answer or uses some mental math strategy to get the answer.

Some children will accomplish this quicker and need less practice, whereas others will need more practice. Thus, don't assign all the exercises in the curriculum by default. Use your judgment, and try to match the amount of exercises to your child's need. The ones that don't get assigned can be used later for review. You can also use games to reinforce the facts, and in place of some of the exercises in the book (a list of online games is provided below).

Learning addition and subtraction facts is very important for later study. For example, we will soon learn regrouping (carrying/borrowing) in addition and in subtraction, which requires the child to be able to recall all the sums of single-digit numbers and corresponding subtraction facts efficiently and fluently.

We will start the chapter by reviewing how to complete the next whole ten. This concept is very important. For example, what number do you add to 23 to get 30? As an equation, we write: $23 + \_\_ = 30$.

In the next lesson, we study sums that go over ten, doing these sums in two parts. For example, in the sum $9 + 7$, the child first completes 10 by adding $9 + 1$. Then, the child adds the rest, or 6, to 10. Learning this prepares the child for addition facts where the sum is more than 10.

The next lessons, *Adding with 9*, *Adding with 8*, *Adding with 7*, and *Adding with 6*, provide lots of practice for learning and memorizing the basic addition facts. There are 20 such facts:

$9 + 2$ till $9 + 9$: 8 facts
$8 + 3$ till $8 + 8$: 6 facts
$7 + 4$ till $7 + 7$: 4 facts
$6 + 5$ till $6 + 6$: 2 facts

After those lessons, we study subtraction. First, the child subtracts to ten. This means subtracting from 14, 15, 16, etc. so that the answer is 10, for example $16 - \_\_ = 10$. In the next step, we study subtractions with an answer less than 10, such as $16 - 7$. The student practices these by subtracting in two parts: First subtracting to ten, then the rest. For example, $16 - 7$ becomes $16 - 6 - 1$, or $14 - 6$ becomes $14 - 4 - 2$. This is a mental math strategy that can be relied on if the child does not know the answer by heart but it is actually not the ultimate goal. Memorizing the basic facts is the preferred way, because it frees up the brain's resources to do other things (such as to think on how to solve a word problem, or how to regroup).

The last part of this chapter includes various lessons titled *Number Rainbows* and *Fact Families with....* These give lots of practice and reinforcement for the basic addition and subtraction facts, emphasizing the connection between addition and subtraction as a strategy for subtraction facts. You can find several helpful videos that match these particular lessons at https://www.mathmammoth.com/videos, in the section for 2nd grade. The lessons also include many word problems.

This chapter includes lots of repetition, drill, and practice. Therefore, you are welcome to mix the lessons from this chapter with some geometry, place value, clock, or measuring, in order to prevent boredom. For example, the child could study geometry and topics of this chapter each day, or study the two different chapters on alternate days. This is not compulsory but just a suggestion to "mix things up" in a somewhat spiral fashion.

Pacing Suggestion for Chapter 3

Please add one day to the pacing for the test if you use it. Note that the lessons in the chapter can take several days to finish. As a general guideline, second graders should finish 1.5-2 pages daily or 8-10 pages a week. Please also see the user guide at https://www.mathmammoth.com/userguides/ .

The Lessons in Chapter 3	page	span	suggested pacing	your pacing
Review: Completing the Next Whole Ten	59	*2 pages*	2 days	
Review: Going Over Ten	61	*2 pages*	1 day	
Adding with 9	63	*2 pages*	1 day	
Adding with 8	65	*2 pages*	1 day	
Adding with 7	67	*2 pages*	1 day	
Adding with 6	69	*2 pages*	1 day	
Review—Facts with 6, 7, and 8	71	*2 pages*	2 days	
Subtract to Ten	73	*2 pages*	1 day	
Difference and How Many More	75	*3 pages*	2 days	
Number Rainbows—11 and 12 (optional)	78	*2 pages*	1 day	
Fact Families with 11	80	*1 page*	1 day	
Fact Families with 12	81	*2 pages*	1 day	
Number Rainbows—13 and 14 (optional)	83	*1 page*	1 day	
Fact Families with 13 and 14	84	*3 pages*	2 days	
Fact Families with 15	87	*2 pages*	1 day	
Fact Families with 16	89	*2 pages*	1 day	
Fact Families with 17 and 18	91	*2 pages*	1 day	
Mixed Review Chapter 3	93	*2 pages*	2 days	
Review Chapter 3	95	*3 pages*	1 day	
Chapter 3 Test (optional)				
TOTALS with optional content		36 pages (39 pages)	22 days (24 days)	

Games and Activities

12 Out (or *11 Out, 13 Out, 14 Out*)

You need: A deck of number cards, or regular playing cards. The values of the face cards are Jack = 11, Queen = 12, King = 13.

Preparation: Choose a target sum, such as 12. The game works best for target sums 14 or less. Deal seven cards to each player. Place the rest face down in a pile in the middle of the table.

Game play: At your turn, first take one card from the pile. Then try to find pairs of cards in your hand that add up to 12, and discard any such pairs. Discard the card 12 (queen) also if you have it. If you cannot find any such pairs, ask for any one card you want (such as 7) from the player to your right (as in "Go Fish"). That player, if he has it, must give it, and you will then discard the pair that makes 12. Then it is the next player's turn. The player who first discards all the cards from his hand is the winner.

Variations:
* Deal more than seven cards.
* Instead of 12, players discard cards that add up to 12, 13, or 14.

Addition (or Subtraction) Challenge

You need: A standard deck of playing cards from which you remove the face cards. For the subtraction challenge, include the face cards also (Jack = 11, Queen = 12, King = 13).

Game Play: In each round, each player is dealt two cards face up, and has to calculate their sum or difference (add/subtract). The player with the highest sum or difference gets all the cards from the other players. After enough rounds have been played to use all of the cards, the player with the most cards wins. If two or more players have the same sum, then those players get an additional two cards and use those to resolve the tie.

Number Bonds in the Pond

You need: A standard deck (or several) of playing cards or number cards. The values of the face cards are Jack = 11, Queen = 12, King = 13.

Preparation: Choose a target sum for the game. If the target sum is 12, make a deck of cards consisting of numbers 1 through 11. If the target sum is 11, make a deck of numbers 1-10. And so on. (The deck always consists of numbers that are from 1 through $X - 1$ where X is the target sum.) Place a target number card face up between the players, and spread out the rest of the cards face down, like a pond, between the players.

Game play: At your turn, if you don't have any cards in your hand, take <u>two</u> cards from the pond. If you do, take <u>one</u> card from the pond. Now check if any two cards in your hand add up to the target number. If so, put those cards away to your personal pile. If not, it is the next player's turn. The game ends when there are no more cards in the pond. The winner is the person with the most cards in their personal pile.

Variation: Allow three cards/numbers to be added to reach the target number.

Note: Depending on the number of players, you may need several decks of cards for the pond.

Get Out of My House

You need: A deck of playing cards or number cards from 3 to 10.

Preparation: On a shared piece of paper, draw boxes (houses) numbered from 6 to 20. This works best as a two-player game, and each player needs seven tokens that are distinct from the other player's tokens. Place the deck of cards in the middle, cards face down.

Game play: During a turn, a player takes two cards from the deck, adds them, and then puts their token in a house with fewer than three of the opponent's tokens. If the house contains one or two of the opponent's tokens, those tokens are given back to the opponent and the player says "Get out of my house." The first player to place all their tokens in houses wins.

Variation: Allow subtraction and/or multiplication to be used, along with addition.

This game is adapted from https://www.earlyfamilymath.org and published here with permission.

Games and Activities at Math Mammoth Practice Zone

Single-Digit Addition
Simple practice of addition facts with single-digit addends.
https://www.mathmammoth.com/practice/addition-single-digit#questions=10&toe=18&pt=general

Hidden Picture Addition Game
Use a number range of 2 to 9 to specifically practice basic addition facts.
https://www.mathmammoth.com/practice/mystery-picture

7 Up Card Game
You will see seven cards dealt face up. Simply choose any two cards that make 10 (or your chosen sum) to discard. When there are no cards that make that sum, click the deck to deal more cards. For this chapter, choose sums of 11, 12, 13, and 14.
https://www.mathmammoth.com/practice/seven-up

Fact Families
Choose which fact family or families to practice, and the program will give you addition and subtraction problems from those, including with missing numbers. For this chapter, choose fact families with 11, 12, 13, 14, and 15.
https://www.mathmammoth.com/practice/fact-families

Mathy's Berry Picking Adventure
Join Mathy, our mammoth mascot, on his berry-picking adventure, and practice your basic addition or subtraction facts!
https://www.mathmammoth.com/practice/mathy-berries#mode=addition-single&duration=2m

https://www.mathmammoth.com/practice/mathy-berries#mode=sub-20&duration=2m

Bingo
Simply click on the right answer in the grid, and it will be colored green. Once you get five in a row, a column, or diagonally, and bingo, you win! For this chapter, choose Addition (Single-Digit) or Subtraction (Under 20).
https://www.mathmammoth.com/practice/bingo

Fruity Math
Click the fruit with the correct answer and try to get as many points as you can within two minutes. The first link below is for addition facts, the second one for subtraction within 0-18.
https://www.mathmammoth.com/practice/fruity-math#op=addition&duration=120&mode=manual&config=2,9x1___3,9x1&max-sum=1000

https://www.mathmammoth.com/practice/fruity-math#op=subtraction&duration=30&mode=manual&config=11,18x1___2,9x1&allow-neg=0

Further Resources on the Internet

We have compiled a list of Internet resources that match the topics in this chapter, including pages that offer:

- **online practice** for concepts;
- online **games**, or occasionally, printable games;
- **animations** and interactive **illustrations** of math concepts;
- **articles** that teach a math concept.

We heartily recommend you take a look! Many of our customers love using these resources to supplement the bookwork. You can use these resources as you see fit for extra practice, to illustrate a concept better and even just for some fun. Enjoy!

Scan me

https://l.mathmammoth.com/gr2ch3

Review: Completing the Next Whole Ten

1. Write the previous and next **whole ten**. Then, circle the ten that is nearer the given number.

a. _____, 56, _____	b. _____, 72, _____	c. _____, 94, _____
d. _____, 37, _____	e. _____, 25, _____	f. _____, 31, _____

> 52 and how many more makes the next ten (60)? We can write 52 + _____ = 60.
> You can solve it using a *helping* problem: <u>2 and how many more makes ten?</u>
> The answer to both problems is the <u>same</u>. It is 8.

2. Complete the next ten. Below, write a helping problem using numbers within 0-10.

a. 17 + ____ = 20	b. 62 + ____ = ____	c. 94 + ____ = ____
7 + ____ = 10	2 + ____ = ____	4 + ____ = ____

3. Complete the next ten. Think of the helping problem that uses numbers within 0-10.

a. 42 + ____ = 50	b. 34 + ____ = ____	c. 66 + ____ = ____
d. 61 + ____ = ____	e. 97 + ____ = ____	f. 83 + ____ = ____

4. Circle the even numbers.   8   9   12   15   10   19   11   6   17

5. Now pick the even numbers from the previous exercise, and write each of them as a double of some number.

a. ____ = ____ + ____	b. ____ = ____ + ____
c. ____ = ____ + ____	d. ____ = ____ + ____

6. Complete the next ten... and then go one more! Compare the two problems in each box.

a. 73 + ____ = 80	b. 35 + ____ = 40	c. 14 + ____ = 20
73 + ____ = 81	35 + ____ = 41	14 + ____ = 21

7. Find your way through the maze! Start at the top. You can only color a square if the sum is a whole ten (10, 20, 30, 40, 50, 60, 70, 80, 90, or 100).

13 + 6	54 + 6	73 + 8	45 + 7	99 + 4
15 + 9	14 + 8	15 + 5	13 + 6	32 + 7
45 + 7	73 + 7	64 + 5	82 + 9	16 + 7
30 + 12	39 + 1	74 + 6	73 + 9	52 + 7
46 + 7	32 + 7	31 + 9	86 + 4	65 + 4
92 + 4	21 + 8	24 + 7	22 + 8	32 + 6
83 + 6	11 + 7	98 + 2	57 + 3	17 + 9
44 + 9	12 + 8	95 + 6	38 + 5	53 + 9
71 + 9	34 + 4	36 + 7	19 + 4	28 + 11
53 + 7	29 + 2	26 + 6	78 + 6	32 + 5

8. Complete the next whole ten. These are more challenging.

a. 17 + ____ + 1 = 20	b. 35 + ____ + 2 = 40	c. 41 + ____ + 6 = 50
12 + ____ + 4 = 20	32 + ____ + 3 = 40	44 + ____ + 3 = 50
13 + ____ + 4 = 20	36 + ____ + 3 = 40	42 + ____ + 5 = 50

9. Find as many different sums as you can to make one hundred!

90 + ____ + ____ = 100	90 + ____ + ____ = 100	90 + ____ + ____ = 100
90 + ____ + ____ = 100	90 + ____ + ____ = 100	90 + ____ + ____ = 100
90 + ____ + ____ = 100	90 + ____ + ____ = 100	90 + ____ + ____ = 100

Review: Going Over Ten

Imagine that 8 wants to get some from 6 in order to make a ten. Six gives two to 8, and has only four left for itself!	Imagine that 9 wants to get some from 7 in order to make a ten. Seven gives one to 9, and has only six left for itself!

8 + 6
  | \\
8 + 2 + 4   In the end, we
                      have 10 and 4.
10 + 4 = 14   We get 14.
	9 + 7
	  | \\
	9 + 1 + 6 In the end, we
	                      have 10 and 6.
	10 + 6 = 16 We get 16.

1. Circle all the blue balls and some of the red ones so that you get a ten. Then add the rest.

a. 8 + 4	b. 9 + 5
10 + _2_ = ____	10 + ____ = ____
c. 8 + 6	d. 9 + 3
10 + ____ = ____	10 + ____ = ____
e. 7 + 5	f. 9 + 8
10 + ____ = ____	10 + ____ = ____

2. Write a number on the empty line inside the balloon so that the numbers in the balloon make a ten. Then add the last number to 10.

a.	b.	c.
(7 + _3_) + 2 = ____	(5 + ___) + 3 = ____	(8 + ___) + 4 = ____
d.	e.	f.
(6 + ___) + 4 = ____	(9 + ___) + 7 = ____	(7 + ___) + 5 = ____

3. Fill in. Imagine that the first number wants to become a ten.

a. 8 + 7	b. 8 + 9	c. 8 + 5
/ \	/ \	/ \
8 + __2__ + __5__	8 + ____ + ____	8 + ____ + ____
10 + __5__ = __15__	10 + ____ = ____	10 + ____ = ____
d. 9 + 4	e. 9 + 6	f. 9 + 9
/ \	/ \	/ \
9 + ____ + ____	9 + ____ + ____	9 + ____ + ____
10 + ____ = ____	10 + ____ = ____	10 + ____ = ____

4. Add so that you get 10, 11, and 12. Notice the patterns!

a.	b.	c.	d.
8 + ____ = 10	7 + ____ = 10	9 + ____ = 10	6 + ____ = 10
8 + ____ = 11	7 + ____ = 11	9 + ____ = 11	6 + ____ = 11
8 + ____ = 12	7 + ____ = 12	9 + ____ = 12	6 + ____ = 12

5. Find the even numbers.

   15   24   58   89   99
   40   51   67   100   2

6. Solve the word problems. ALSO, write an addition & subtraction sentence for them!

a. You have $8 and you buy a toy for $5 and candy for $2. How much money do you have now?
b. Lucy had $8. Then she found $5 in her piggy bank, and her mom gave her $2. How much money does she have now?
c. Matthew had $8. He spent $3 on a bottle of juice. Later he found $2 in the street. How much money does he have now?

Adding with 9

Imagine that 9 *really* wants to be a 10! It takes one from the other number (from 5). So, 9 becomes 10, and four dots are left over.

9 + 5 = 10 + 4 = 14

9 wants to be a 10! So, it takes one from the other number (from 3). So, 9 becomes 10, and two dots are left over.

9 + 3 = 10 + 2 = 12

Use the list on the right to practice. Don't write the answers there.
Just point to different problems and say the answer aloud.

1. Circle the ten, then add.

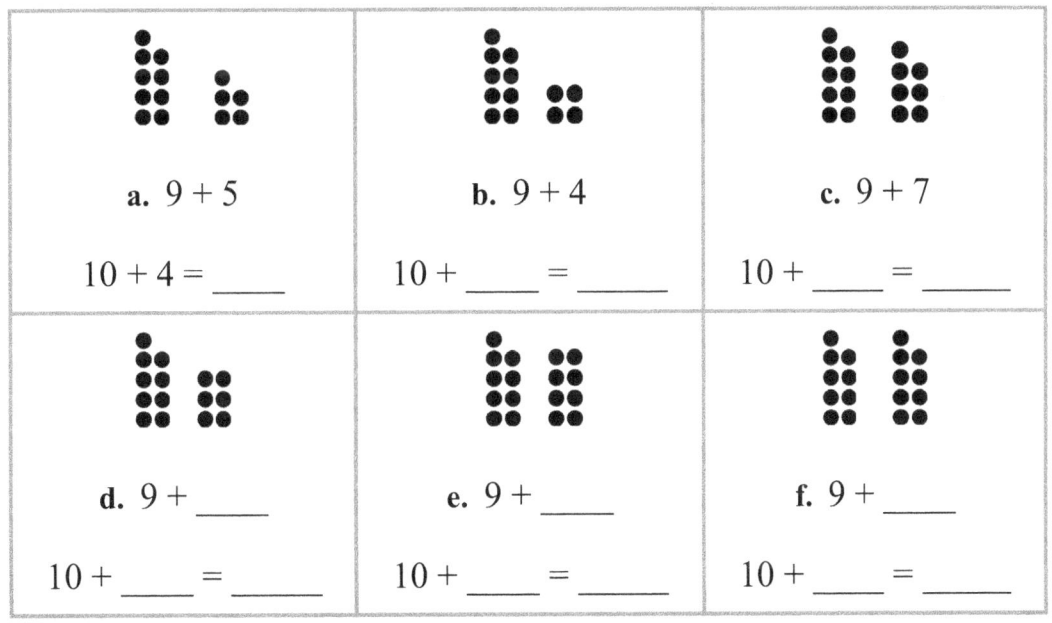

a. 9 + 5
10 + 4 = _____

b. 9 + 4
10 + ___ = _____

c. 9 + 7
10 + ___ = _____

d. 9 + ___
10 + ___ = _____

e. 9 + ___
10 + ___ = _____

f. 9 + ___
10 + ___ = _____

9 + 1 = ☐
9 + 2 = ☐
9 + 3 = ☐
9 + 4 = ☐
9 + 5 = ☐
9 + 6 = ☐
9 + 7 = ☐
9 + 8 = ☐
9 + 9 = ☐

2. It is good to memorize the doubles, also. Fill in.

2 + 2 = _____	5 + 5 = _____	8 + 8 = _____
3 + 3 = _____	6 + 6 = _____	9 + 9 = _____
4 + 4 = _____	7 + 7 = _____	10 + 10 = _____

3. Add to nine. Think how 9 wants to be a ten, and takes 1 from the other number.

a. 9 + 6 10 + 5 = _____	b. 9 + 8 10 + _____ = _____	c. 9 + 5 10 + _____ = _____
d. 9 + 7 10 + _____ = _____	e. 9 + 9 10 + _____ = _____	f. 9 + 3 10 + _____ = _____

4. Practice the facts with nine. Don't write the answers down; just practice the sums.

$9 + 0 = \square$ $9 + 5 = \square$ $9 + 9 = \square$
$9 + 3 = \square$ $9 + 6 = \square$ $9 + 1 = \square$ $9 + 4 = \square$
$9 + 7 = \square$ $9 + 8 = \square$ $9 + 2 = \square$ $9 + 10 = \square$

5. Add. Remember, you can add both ways. For example, 7 + 9 is the same as 9 + 7.

a. 9 + 4 = _____ 8 + 9 = _____ 9 + 5 = _____	b. 9 + 7 = _____ 4 + 9 = _____ 9 + 4 = _____	c. 3 + 9 = _____ 9 + 2 = _____ 9 + 9 = _____	d. 5 + 9 = _____ 8 + 9 = _____ 9 + 6 = _____

6. What is missing?

a. 9 + ☐ = 13 9 + ☐ = 15	b. 9 + ☐ = 16 9 + ☐ = 14	c. ☐ + 9 = 17 ☐ + 9 = 11

Puzzle Corner

You can use this same "trick" with 19, 29, 39, 49, and so on. Imagine that 49 *really* wants to be 50, and so it "takes" 1 from the other number. Solve.

a. 49 + 7 = _____   b. 59 + 5 = _____   c. 69 + 3 = _____

   19 + 6 = _____   89 + 9 = _____   29 + 6 = _____

Adding with 8

Imagine that 8 wants to be a 10! It takes two from the other number (from 3). So, 8 becomes 10, and only 1 is left over.

8 + 3 = 10 + 1 = 11

8 wants to be a 10! So, it takes two from the other number (from 5). So, 8 becomes 10, and 3 are left over.

8 + 5 = 10 + 3 = 13

Use the list on the right to practice. Don't write the answers there. Just point to different problems and say the answer aloud.

1. Add. First, circle the ten.

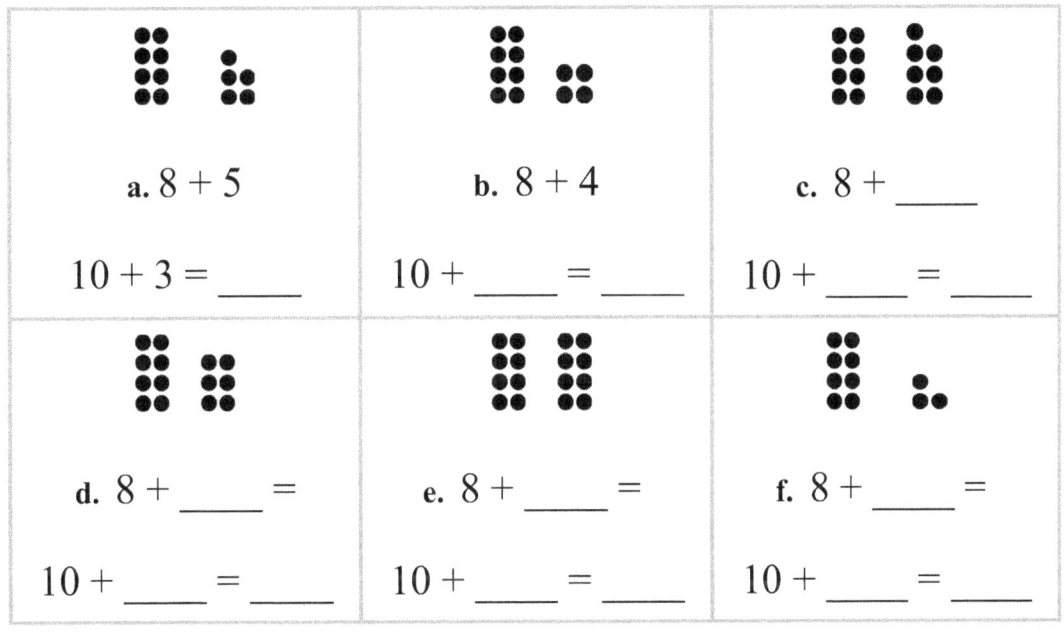

a. 8 + 5
10 + 3 = ___

b. 8 + 4
10 + ___ = ___

c. 8 + ___
10 + ___ = ___

d. 8 + ___ =
10 + ___ = ___

e. 8 + ___ =
10 + ___ = ___

f. 8 + ___ =
10 + ___ = ___

8 + 1 = ☐
8 + 2 = ☐
8 + 3 = ☐
8 + 4 = ☐
8 + 5 = ☐
8 + 6 = ☐
8 + 7 = ☐
8 + 8 = ☐
8 + 9 = ☐

2. It is good to memorize the doubles, also. Fill in.

2 + 2 = ___	5 + 5 = ___	8 + 8 = ___
3 + 3 = ___	6 + 6 = ___	9 + 9 = ___
4 + 4 = ___	7 + 7 = ___	10 + 10 = ___

Addition facts with eight. Do not write the answers down, but just practice the sums.

8 + 0 = ☐	8 + 5 = ☐	8 + 8 = ☐	8 + 9 = ☐
8 + 3 = ☐	8 + 7 = ☐	8 + 1 = ☐	8 + 4 = ☐
8 + 10 = ☐	8 + 1 = ☐	8 + 6 = ☐	8 + 2 = ☐

3. Add and fill in what is missing.

a. 8 + 4 = ____	**b.** 7 + 8 = ____	**c.** 3 + 8 = ____
8 + 6 = ____	8 + 5 = ____	8 + 9 = ____
d. 8 + ____ = 13	**e.** 8 + ____ = 12	**f.** ____ + 8 = 11
8 + ____ = 15	8 + ____ = 16	____ + 8 = 14

4. **a.** Jenny ate 8 strawberries, and Jack ate 5 more than what Jenny did. How many strawberries did Jack eat?

   b. Ashley is 13 years old, and Maryann is 5. How many years older is Ashley than Maryann?

5. Find the patterns and continue them.

a. 8 + 2 = ____	**b.** 18 + 2 = ____	**c.** $\frac{1}{2}$ of 0 is ____.
8 + 4 = ____	18 + 4 = ____	$\frac{1}{2}$ of 2 is ____.
8 + 6 = ____	18 + 6 = ____	$\frac{1}{2}$ of 4 is ____.
8 + ___ = ____	18 + ___ = ____	$\frac{1}{2}$ of ____ is ____.
___ + ___ = ____	___ + ___ = ____	$\frac{1}{2}$ of ____ is ____.
___ + ___ = ____	___ + ___ = ____	$\frac{1}{2}$ of ____ is ____.
___ + ___ = ____	___ + ___ = ____	$\frac{1}{2}$ of ____ is ____.

Adding with 7

We have already studied these facts:

7 + 8 = ____   8 + 7 = ____

7 + 9 = ____   9 + 7 = ____

7 + 10 = ____   10 + 7 = ____

These are the new facts with 7:

7 + 4 = ____   7 + 6 = ____

7 + 5 = ____   7 + 7 = ____

Tricks for remembering facts with 7

- 7 + 7 = 14 is one of the doubles. Memorize all the doubles! But if you forget, you can do 5 + 5 = 10, then 6 + 6 = 12, and *then* 7 + 7 = 14.

- 7 + 6 is *just one more* than the doubles fact 6 + 6 = 12. So, it is 13. Or, 7 + 6 is *just one less* than the doubles fact 7 + 7 = 14.

- 7 + 4 is *just one more* than the ten-fact 7 + 3 = 10. So, 7 + 4 is 11.

- 7 + 5 is just one more than 7 + 4, or just one less than 7 + 6, so if you remember those, you can figure out 7 + 5, too. Or maybe you have your own trick for it!

7 + 1 = ☐
7 + 2 = ☐
7 + 3 = ☐
7 + 4 = ☐
7 + 5 = ☐
7 + 6 = ☐
7 + 7 = ☐
7 + 8 = ☐
7 + 9 = ☐

Use the list on the right to practice. Don't write the answers there. Just point to different problems and say the answer aloud.

1. Let's practice doubles—and doubles plus **one more**.
   Notice: the answer is also just one more!

a. 6 + 6 = ____ 6 + 7 = ____	**b.** 7 + 7 = ____ 7 + 8 = ____	**c.** 8 + 8 = ____ 8 + 9 = ____
d. 9 + 9 = ____ 9 + 10 = ____	**e.** 5 + 5 = ____ 6 + 5 = ____	**f.** 4 + 4 = ____ 4 + 5 = ____

Addition facts with seven. Do not write the answers down, but just practice the sums.

7 + 0 = ☐	7 + 5 = ☐	7 + 6 = ☐	7 + 9 = ☐
7 + 3 = ☐	7 + 9 = ☐	7 + 7 = ☐	7 + 4 = ☐
7 + 10 = ☐	7 + 8 = ☐	7 + 1 = ☐	7 + 2 = ☐

2. Fill in the missing numbers.

a. 7 + 4 = ___ 6 + 7 = ___ 7 + 5 = ___	**b.** 8 + 7 = ___ 7 + 10 = ___ 3 + 7 = ___	**c.** 7 + ___ = 14 7 + ___ = 13 7 + ___ = 15	**d.** 7 + ___ = 12 7 + ___ = 16 7 + ___ = 11
e. 7 + 7 = ___ 9 + 7 = ___ 7 + 8 = ___	**f.** 4 + 7 = ___ 7 + 9 = ___ 3 + 7 = ___	**g.** 8 + ___ = 13 8 + ___ = 16 8 + ___ = 17	**h.** ___ + 7 = 17 ___ + 7 = 10 ___ + 7 = 12

3. Try these boxes!

Add 7 each time.
Add 8 each time.
Add 9 each time.

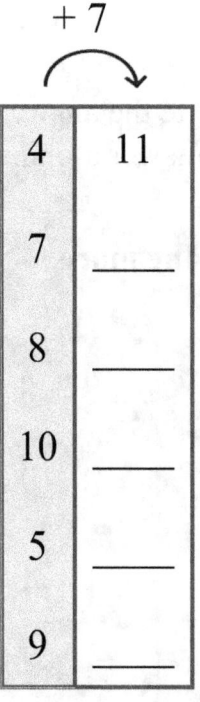
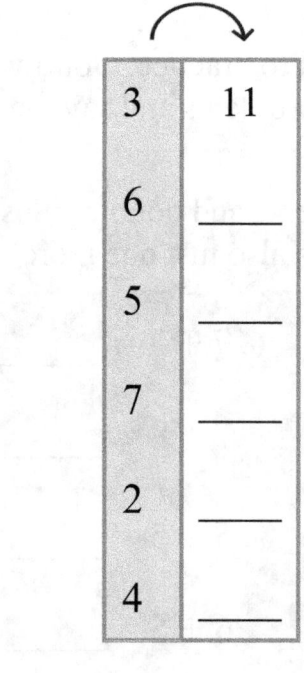
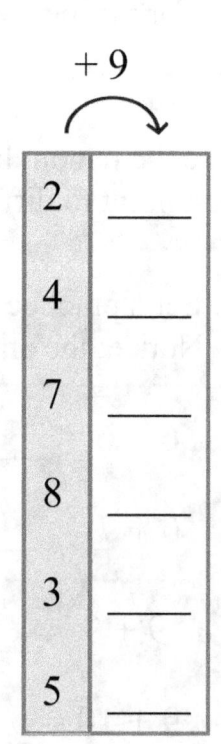

Adding with 6

6 + 5 = ___	6 + 6 = ___
This is *just one more* than 5 + 5 = 10.	This is one of the doubles!

Here are addition facts where we add to six. Do not write the answers down. Just go over the problems until you remember them easily.

6 + 0 = ☐   6 + 5 = ☐   6 + 9 = ☐   6 + 6 = ☐
6 + 3 = ☐   6 + 7 = ☐   6 + 4 = ☐
6 + 10 = ☐   6 + 1 = ☐   6 + 2 = ☐   6 + 8 = ☐

1. Fill in the missing numbers.

a.	b.	c.	d.
6 + 4 = ___	6 + 8 = ___	6 + ___ = 14	___ + 6 = 12
6 + 6 = ___	6 + 9 = ___	6 + ___ = 16	___ + 6 = 15
6 + 5 = ___	6 + 7 = ___	6 + ___ = 12	___ + 6 = 11

e.	f.	g.	h.
5 + 6 = ___	9 + 6 = ___	7 + ___ = 14	___ + 6 = 13
6 + 7 = ___	8 + 6 = ___	8 + ___ = 14	___ + 6 = 14
4 + 6 = ___	6 + 6 = ___	9 + ___ = 14	___ + 6 = 15

> **Trick!** When you add three or four numbers, first add the numbers that make ten. It makes adding easier!
>
> 8 + <u>6</u> + <u>4</u>   <u>5</u> + 3 + 2 + <u>5</u>
>
> = 8 + 10 = 18   = 10 + 5 = 15

2. Add. *First* find the numbers that make 10. You can circle or color them. Then add the rest. This is like hide-and-seek! Where are those numbers hiding that make ten?

a.	b.	c.
1 + 6 + 9 = ___	3 + 6 + 7 + 2 = ___	6 + 5 + 1 + 4 = ___
6 + 8 + 2 = ___	1 + 5 + 5 + 7 = ___	8 + 3 + 2 + 6 = ___
5 + 7 + 5 = ___	2 + 7 + 8 + 2 = ___	9 + 6 + 1 + 4 = ___

3. Solve the word problems.

   a. There were some apples on the table. Children came in and ate five apples. Later, Mom saw seven apples left on the table. How many apples had there been at first?

   b. Jeremy had $12. He bought a toy truck, and then he had $6 left. How much did the toy truck cost?

   c. Mom bought a bunch of bananas. She ate one, Dad ate two, and the children ate two. Then there were four bananas left. How many bananas did Mom buy?

   d. Mike solved 9 math problems. Scott solved 5 more than Mike. How many did Scott solve?

   e. Elena solved 14 math problems and Ashley solved 7. How many more did Elena solve than Ashley?

Review—Facts with 6, 7, and 8

1. Here are the 20 addition facts with single-digit numbers where the sum is between 10 and 20. Connect the problems to the right answer.

6 + 6		8 + 6		
5 + 8		5 + 7		9 + 9
	11		**15**	7 + 9
9 + 5		9 + 2		
	12		**16**	8 + 7
5 + 6		4 + 7		
	13		**17**	9 + 8
3 + 9		9 + 4		
	14		**18**	8 + 8
7 + 7		6 + 7		
				6 + 9
8 + 3		4 + 8		

2. Figure out the pattern and continue it.

a.	b.	c.
9 + ___ = 19	___ + 16 = 17	6 + ___ = 6
8 + ___ = 18	___ + 14 = 17	6 + ___ = 8
7 + ___ = 17	___ + 12 = 17	6 + ___ = 10
___ + ___ = ___	___ + ___ = ___	___ + ___ = ___
___ + ___ = ___	___ + ___ = ___	___ + ___ = ___
___ + ___ = ___	___ + ___ = ___	___ + ___ = ___
___ + ___ = ___	___ + ___ = ___	___ + ___ = ___
___ + ___ = ___	___ + ___ = ___	___ + ___ = ___

3. Fill in the addition table.

+	6	8	4	5	7	3	9
7							
9							
5							

4. Solve.

a. A herd of elephants was feeding on the grass. Four of them left, but fourteen stayed feeding. How many elephants are in the herd?

b. Sarah has five more dolls than Annie. Sarah has 10 dolls. How many does Annie have?
Hint 1: Draw Sarah's dolls. Hint 2: Think which girl has more dolls.
Should you draw more or fewer dolls for Annie?

c. Ronnie and Luis emptied waste baskets. Ronnie emptied four more waste baskets than Luis. Luis emptied five baskets. How many did Ronnie empty?
Hint 1: Draw Luis's baskets. Hint 2: Think which boy emptied more of them.
Should you draw more or fewer baskets for Ronnie?

d. Cynthia ate 10 peanuts. Marie ate 7 more than Cynthia. How many did Marie eat?

5. Add. In some problems, you can find numbers that *make a ten*.

a.	b.	c.
6 + 6 + 2 = _____	8 + 6 + 3 = _____	6 + 2 + 3 + 7 = _____
1 + 4 + 9 = _____	2 + 2 + 8 = _____	3 + 6 + 7 + 2 = _____

Subtract to Ten

1. Subtract the "ones" that are not in the ten-group. You should only have *ten* left!

a. 14 − ___ = 10	b. 16 − ___ = ___	c. 15 − ___ = ___
d. 13 − ___ = 10	e. 17 − ___ = ___	f. 19 − ___ = ___

Subtracting in parts

Let's subtract 13 − 5. First we subtract enough dots that we have only 10 left. So, first we take away 3 dots. 13 − 3 = 10.

We still need to subtract 2 more. We subtract those from 10. There are 8 left.

13 − 5
/ \
13 − 3 − 2

= 8

2. First subtract enough that you have only 10 left. Then subtract the rest. You can cover some dots to help.

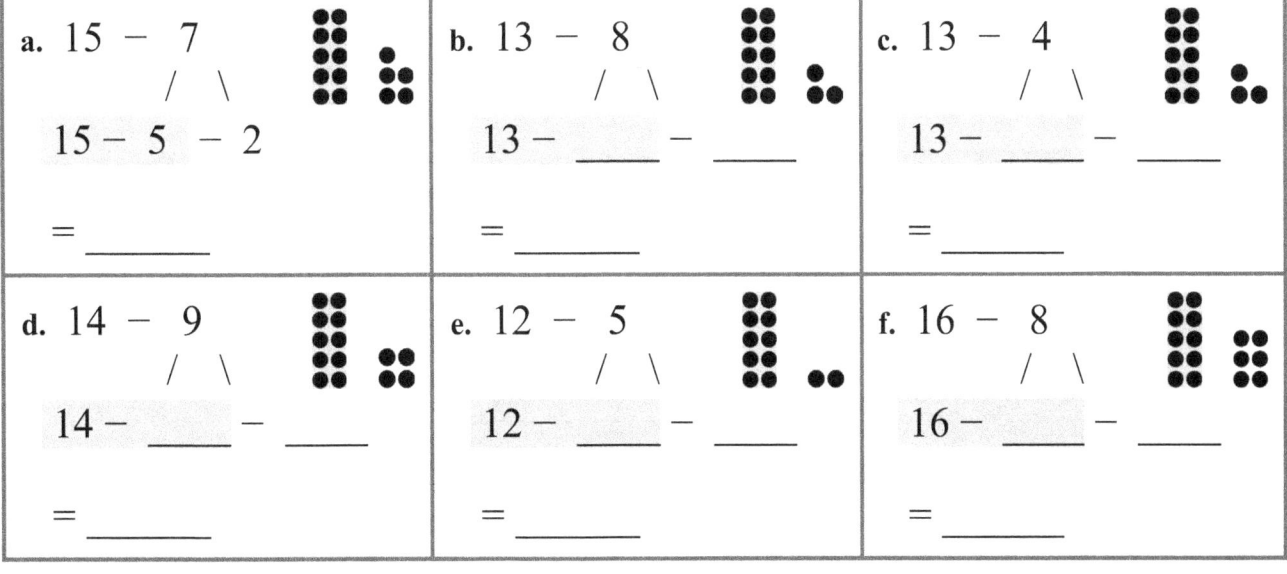

a. 15 − 7
    / \
15 − 5 − 2

= ___

b. 13 − 8
    / \
13 − ___ − ___

= ___

c. 13 − 4
    / \
13 − ___ − ___

= ___

d. 14 − 9
    / \
14 − ___ − ___

= ___

e. 12 − 5
    / \
12 − ___ − ___

= ___

f. 16 − 8
    / \
16 − ___ − ___

= ___

3. First subtract enough that you have only 10 left. Then subtract the rest.

a. 16 − 7
      / \
 16 − ___ − ___
 	
 = ___
	b. 12 − 4
	      / \
	 12 − ___ − ___
	
	 = ___
		c. 13 − 6
		      / \
		 13 − ___ − ___
		
		 = ___

d. 11 − 3
      / \
 11 − ___ − ___
 	
 = ___
	e. 12 − 7
	      / \
	 12 − ___ − ___
	
	 = ___
		f. 15 − 8
		      / \
		 15 − ___ − ___
		
		 = ___

4. Subtract. You can cover dots to help.

a.
12 − 4 = ___
12 − 5 = ___
12 − 3 = ___
12 − 6 = ___

b.
15 − 6 = ___
15 − 9 = ___
15 − 7 = ___
15 − 8 = ___

c.
14 − 5 = ___
14 − 8 = ___
14 − 7 = ___
14 − 6 = ___

5. First subtract those that are not in the ten-group. Compare the top and bottom problems.

a.
15 − 7 = ___
25 − 7 = ___

b.
13 − 6 = ___
23 − 6 = ___

c.
16 − 9 = ___
26 − 9 = ___

Puzzle Corner Can you apply the idea of this lesson to larger numbers? First, subtract to the previous whole ten. Then, subtract some more.

a. 22 − 7
    / \
22 − ___ − ___ = ___

b. 34 − 5
    / \
34 − ___ − ___ = ___

c. 72 − 6
    / \
72 − ___ − ___ = ___

Difference and How Many More

The difference or distance between two numbers means how far apart they are from each other on the number line. The difference between 3 and 12 is 9, because they are NINE steps apart.

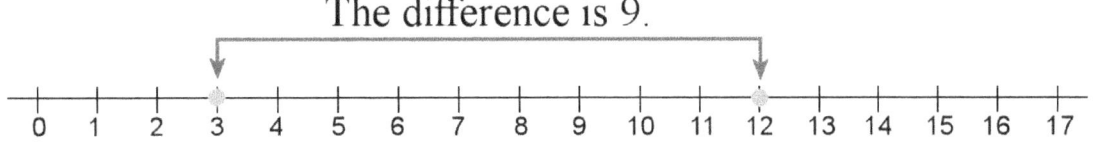

1. Find the differences between these numbers using the number line above.

   a. difference between 10 and 6: _____
   b. difference between 12 and 8: _____

   c. difference between 14 and 2: _____
   d. difference between 17 and 6: _____

We can solve the difference between two numbers by **subtracting**.
What is the difference between 10 and 4? Subtract $10 - 4 = 6$. The difference is 6.

2. Write a subtraction to find the difference between the numbers.

a. The difference between 10 and 4	b. The difference between 2 and 9	c. The difference between 8 and 3
___ − ___ = ___	___ − ___ = ___	___ − ___ = ___
d. The difference between 20 and 50	e. The difference between 10 and 90	f. The difference between 19 and 8
___ − ___ = ___	___ − ___ = ___	___ − ___ = ___

3. Solve the subtractions by thinking of the <u>distance between the numbers</u>—how far apart they are from each other.

a. $20 - 16 =$ ___	b. $40 - 38 =$ ___	c. $65 - 61 =$ ___	d. $36 - 31 =$ ___
e. $100 - 99 =$ ___	f. $87 - 84 =$ ___	g. $55 - 50 =$ ___	h. $79 - 78 =$ ___

> You can also solve the difference between two numbers by thinking of addition: how many more do you need to add to the one number to get the other?
>
> For example, to find the difference between 12 and 7, think: $7 + \underline{} = 12$. ("7 and how many more makes 12?") The answer is 5.

4. Write a *"how many more"* addition to find the difference between the numbers.

a. The difference between 10 and 6	**b.** The difference between 6 and 12
$6 + \underline{} = 10$	$6 + \underline{} = 12$
c. The difference between 15 and 8	**d.** The difference between 4 and 11
$\underline{} + \underline{} = \underline{}$	$\underline{} + \underline{} = \underline{}$

5. Subtract. Think how far apart the two numbers are from each other.

a. $15 - 12 = \underline{}$ (+3)	**b.** $11 - 9 = \underline{}$	**c.** $16 - 11 = \underline{}$
12 and *how many more* makes 15?	9 and *how many more* makes 11?	11 and *how many more* makes 16?

There are two ways to find a difference between two numbers:	
(1) Subtraction	**(2) A *"how many more"* addition**
Find the difference between 100 and 2. It is easier to subtract $100 - 2 = 98$. The difference is 98.	Find the difference between 100 and 95. It is easier to think: $95 + \underline{} = 100$. The difference is 5.

6. Find the differences.

a. The difference between 60 and 56	**b.** The difference between 22 and 20
c. The difference between 35 and 1	**d.** The difference between 67 and 3
e. The difference between 50 and 30	**f.** The difference between 40 and 100

> Whenever a word problem asks "*how many more,*" you can solve it in two ways.
> You can either subtract, or you can write a "*how many more*" addition.
> Either way, you are finding the difference between the two numbers.

7. Solve the word problems.

a. Jane is on page 20 and Boyd is on page 17 of the same book.
How many more pages has Jane read?

b. Mom has one dozen eggs plus five in another carton. A dozen means 12.
How many eggs does Mom have?

c. Barb is reading a 50-page book. She is on page 42.
How many more pages does she have left to read?

d. Janet worked in the garden for 2 hours in the morning and 3 hours in the afternoon. Andy worked for 8 hours in the shop.
Who worked more hours?

How many more?

e. Betty is going batty with flies! She killed 28 flies. Her husband killed 5 flies.
How many more did she kill than him?

f. The next day, Betty was again going batty with flies. She killed 5 flies in the living room, 12 in the kitchen, and 2 in her room.
How many flies did she kill in total?

g. Matthew had $12 and Bob had $6. Then both brothers worked helping Dad in the garden. Matthew earned $5 and Bob earned $9.
Now, who has more money?

How much more?

Number Rainbows—11 and 12

This is a number rainbow for 11. If two numbers are connected with an arc, they add up to 11. Use the number rainbow to help you with addition and subtraction facts!

1. Practice subtraction from 11. Don't write the answers; just think them in your head.

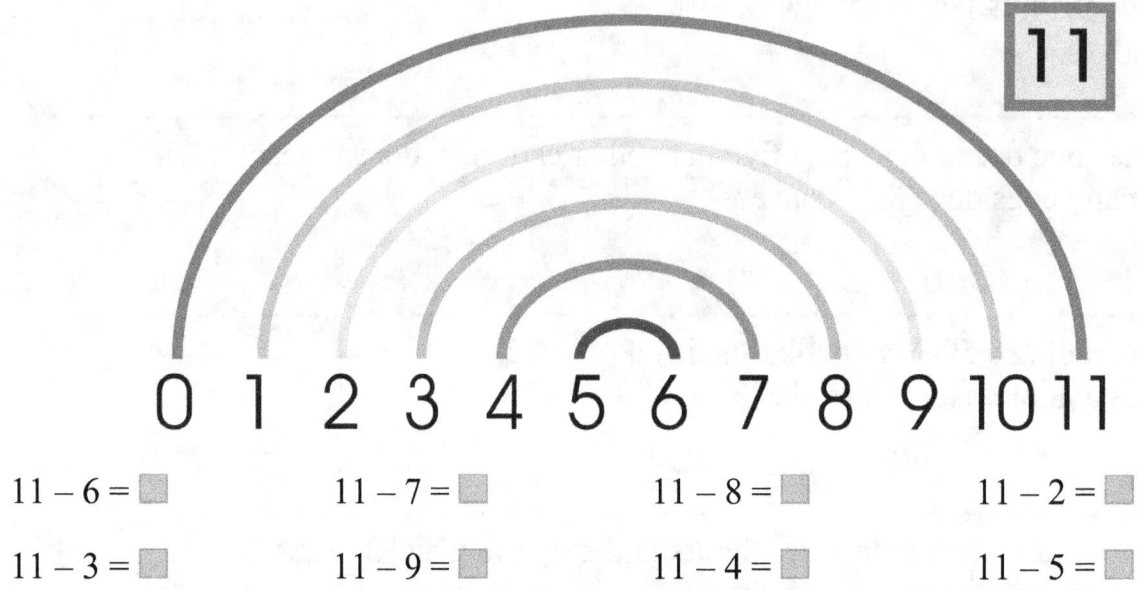

11 − 6 = ☐   11 − 7 = ☐   11 − 8 = ☐   11 − 2 = ☐

11 − 3 = ☐   11 − 9 = ☐   11 − 4 = ☐   11 − 5 = ☐

2. Similarly, practice subtraction from 12.

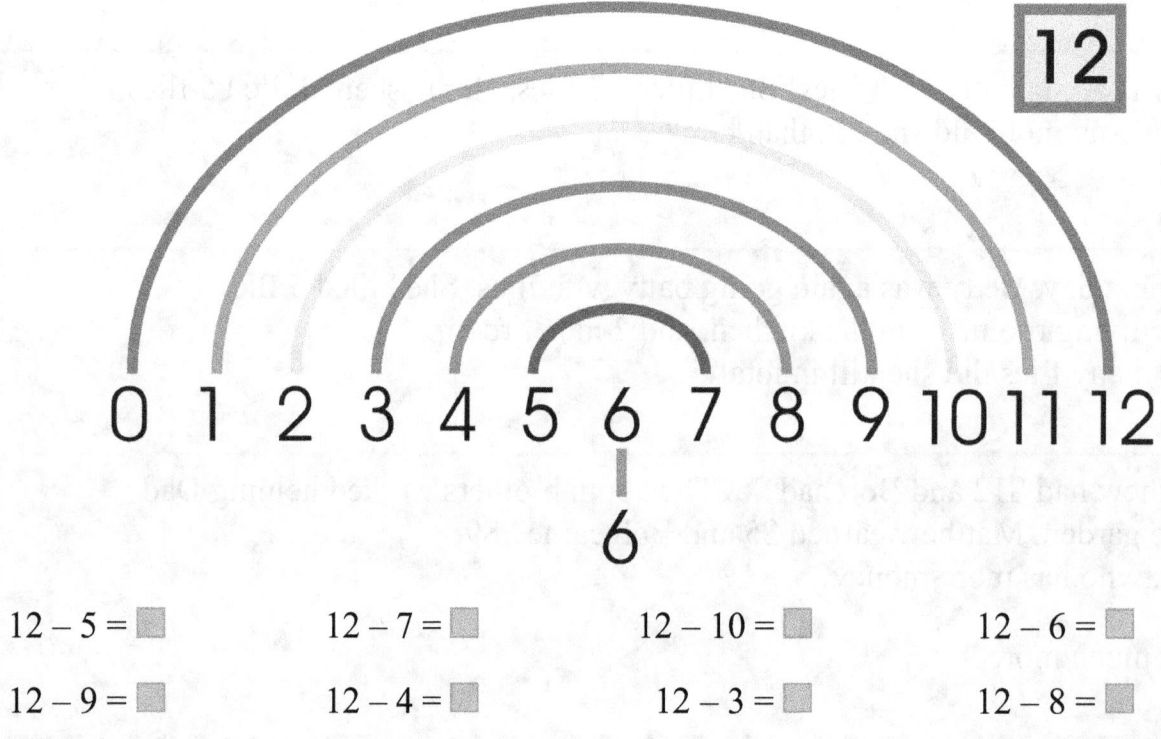

12 − 5 = ☐   12 − 7 = ☐   12 − 10 = ☐   12 − 6 = ☐

12 − 9 = ☐   12 − 4 = ☐   12 − 3 = ☐   12 − 8 = ☐

3. Fill and color the number rainbows. Don't look at the previous page!
   Then practice the subtraction problems.

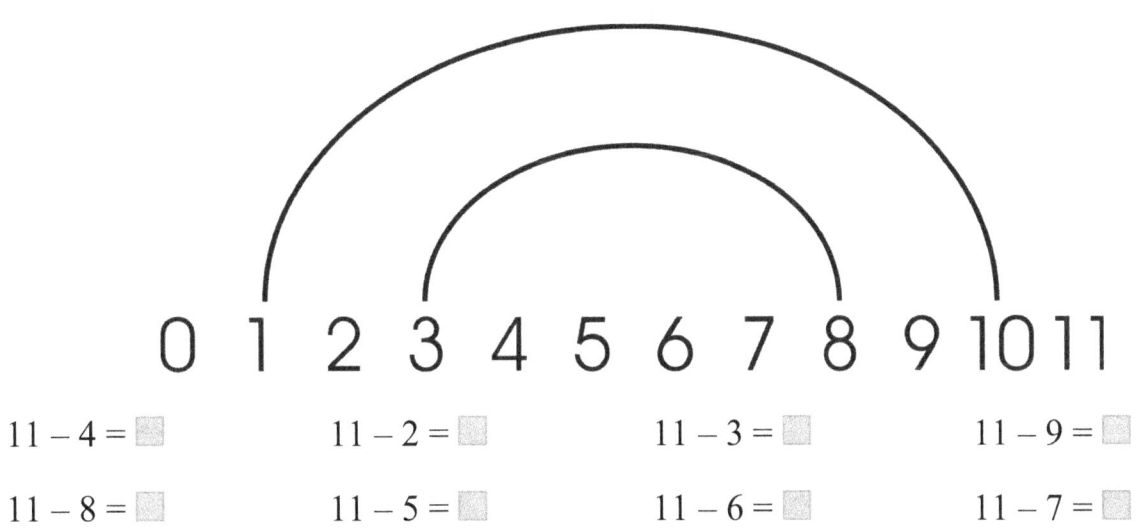

11 − 4 =	11 − 2 =	11 − 3 =	11 − 9 =
11 − 8 =	11 − 5 =	11 − 6 =	11 − 7 =

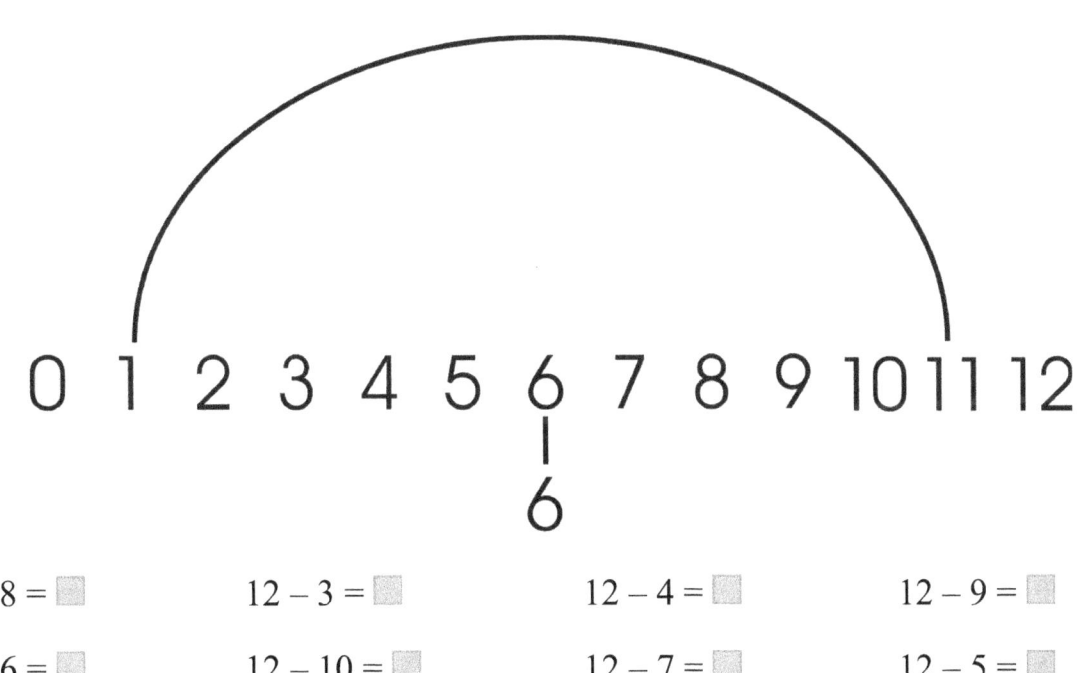

12 − 8 =	12 − 3 =	12 − 4 =	12 − 9 =
12 − 6 =	12 − 10 =	12 − 7 =	12 − 5 =

For more practice, make your own number rainbows and subtractions on blank paper!

Fact Families with 11

1. Fill in. In each fact family, color enough marbles to equal the first number. Then use another color to color the rest.

Fact families with 11		
10, 1, and 11	10 + 1 = ____	11 − 10 = ____
	1 + 10 = ____	11 − 1 = ____
9, ____, and 11	9 + ____ = 11	____ − ____ = ____
	____ + ____ = ____	____ − ____ = ____
8, ____, and 11	8 + ____ = 11	____ − ____ = ____
	____ + ____ = ____	____ − ____ = ____
7, ____, and 11	7 + ____ = 11	____ − ____ = ____
	____ + ____ = ____	____ − ____ = ____
6, ____, and 11	6 + ____ = 11	____ − ____ = ____
	____ + ____ = ____	____ − ____ = ____

2. Check yourself! Can you subtract quickly without looking above?

a. 11 − 10 = ____	**b.** 11 − 2 = ____	**c.** 11 − 3 = ____
11 − 9 = ____	11 − 4 = ____	11 − 6 = ____
11 − 6 = ____	11 − 5 = ____	11 − 9 = ____
11 − 8 = ____	11 − 7 = ____	11 − 4 = ____

Fact Families with 12

1. Fill in. In each fact family, color enough marbles to equal the first number. Then use another color to color the rest.

Fact families with 12		
10, 2, and 12	10 + 2 = ____	12 − 10 = ____
	2 + 10 = ____	12 − 2 = ____
9, ____, and 12	9 + ____ = 12	____ − ____ = ____
	____ + ____ = ____	____ − ____ = ____
8, ____, and 12	8 + ____ = 12	____ − ____ = ____
	____ + ____ = ____	____ − ____ = ____
7, ____, and 12	____ + ____ = ____	____ − ____ = ____
	____ + ____ = ____	____ − ____ = ____
6, ____, and 12	____ + ____ = ____	____ − ____ = ____

2. Check! Can you subtract quickly from 12 and from 11 without looking above?

a.	b.	c.	d.
12 − 4 = ____	11 − 8 = ____	12 − 6 = ____	12 − 3 = ____
11 − 9 = ____	12 − 7 = ____	11 − 4 = ____	12 − 10 = ____
12 − 8 = ____	11 − 3 = ____	12 − 9 = ____	11 − 5 = ____
11 − 6 = ____	12 − 5 = ____	12 − 4 = ____	11 − 7 = ____

3. Practice "how many more" additions! Remember the fact families with 11 and 12.

a. 6 + ___ = 11	b. 7 + ___ = 12	c. ___ + 9 = 11	d. ___ + 6 = 12
8 + ___ = 11	8 + ___ = 12	___ + 7 = 11	___ + 9 = 12

4. *Explain* how you can use <u>addition</u> to solve a subtraction problem, such as 11 − 8.

5. Find the pattern and continue it.

a. 16 − 1 = ___	b. 0 + 17 = ___	c. 15 − 1 = ___
16 − 3 = ___	2 + 15 = ___	15 − 3 = ___
16 − 5 = ___	4 + 13 = ___	15 − 5 = ___
___ − ___ = ___	___ + ___ = ___	___ − ___ = ___
___ − ___ = ___	___ + ___ = ___	___ − ___ = ___
___ − ___ = ___	___ + ___ = ___	___ − ___ = ___
___ − ___ = ___	___ + ___ = ___	___ − ___ = ___
___ − ___ = ___	___ + ___ = ___	___ − ___ = ___

Puzzle Corner Imagine 14 baby blocks in three stacks. One stack has 6 and the third stack has 4. How many are in the middle stack?

We can write an addition where one number is missing: 6 + ___ + 4 = 14.
Figure out a way to solve this problem! Then solve the rest of the problems below.

a. 6 + ___ + 4 = 14	b. 2 + ___ + 2 = 8	c. 10 + ___ + 4 = 17
8 + ___ + 3 = 13	3 + ___ + 3 = 9	10 + ___ + 2 = 15

Number Rainbows—13 and 14

1. Fill in and color the number rainbows. Then practice the subtractions.

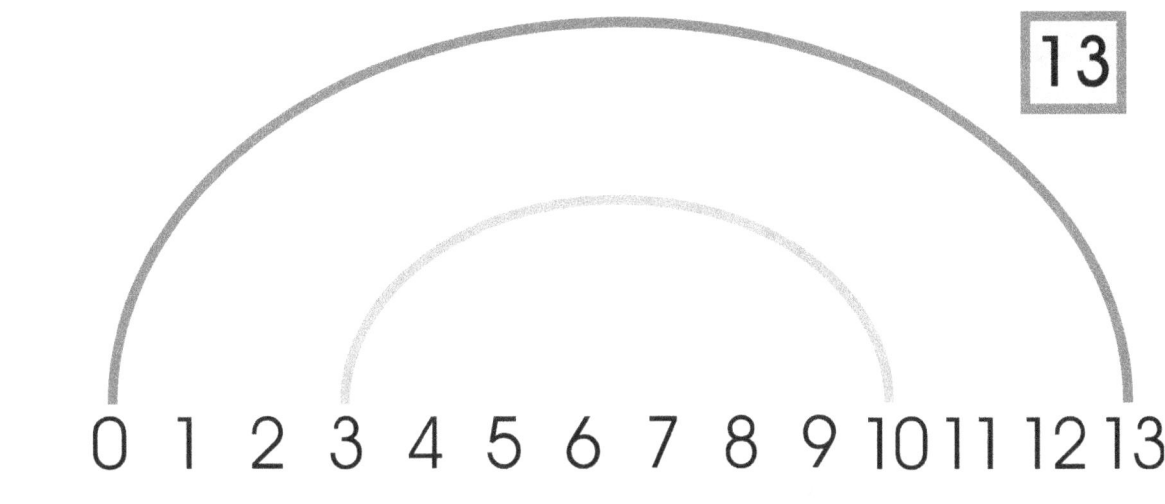

13 − 7 = ☐   13 − 4 = ☐   13 − 9 = ☐   13 − 10 = ☐

13 − 5 = ☐   13 − 6 = ☐   13 − 11 = ☐   13 − 8 = ☐

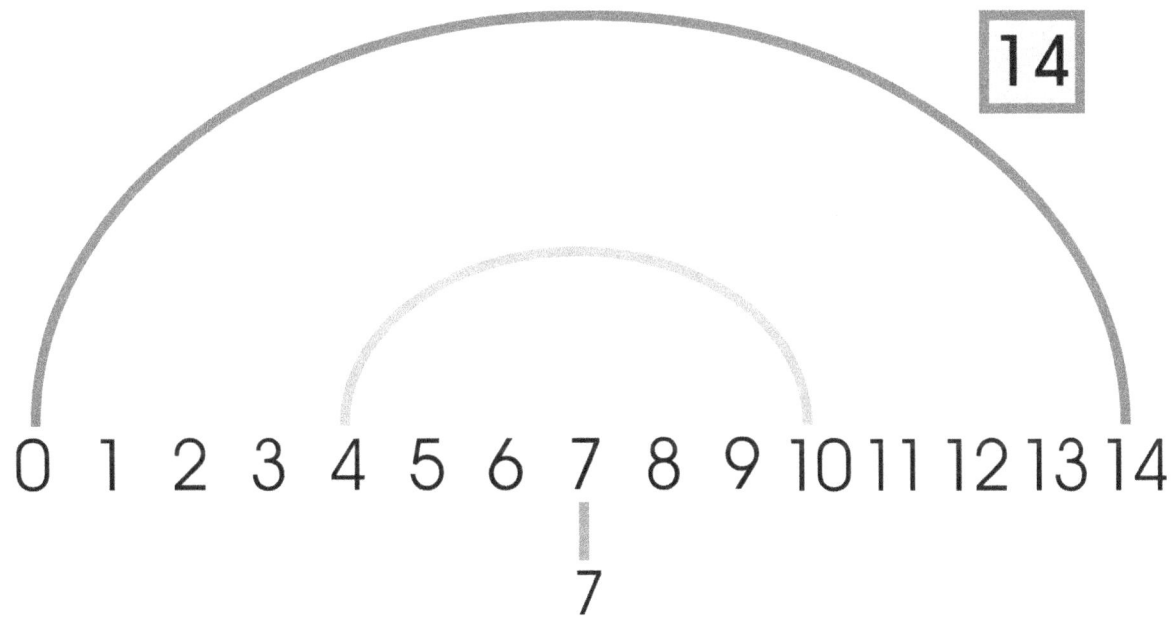

14 − 8 = ☐   14 − 3 = ☐   14 − 7 = ☐   14 − 6 = ☐

14 − 5 = ☐   14 − 9 = ☐   14 − 11 = ☐   14 − 4 = ☐

For more practice, make your own number rainbows and subtractions on blank paper!

Fact Families - 13 and 14

1. Fill in. In each fact family, color the marbles so they match the numbers in it.

Fact families with 13		
10, 3, and 13	10 + 3 = ____	13 − 10 = ____
	3 + 10 = ____	13 − 3 = ____
9, ____, and 13	9 + ____ = 13	____ − ___ = ____
	____ + ___ = ____	____ − ___ = ____
8, ____, and 13	8 + ____ = 13	____ − ___ = ____
	____ + ___ = ____	____ − ___ = ____
7, ____, and 13	7 + ____ = 13	____ − ___ = ____
	____ + ___ = ____	____ − ___ = ____

2. Connect with a line the problems that are from the same fact family. You don't need to write the answers.

13 − 7 = ▪	11 − 4 = ▪	12 − 7 = ▪
5 + ▪ = 12	11 − 8 = ▪	13 − 6 = ▪
11 − 3 = ▪	5 + ▪ = 13	3 + ▪ = 12
8 + ▪ = 13	12 − 5 = ▪	13 − 5 = ▪
12 − 3 = ▪	6 + ▪ = 13	3 + ▪ = 11
7 + ▪ = 11	9 + ▪ = 12	4 + ▪ = 11

3. Fill in. In each fact family, color the marbles so they match the numbers in it.

Fact families with 14		
10, 4, and 14	10 + 4 = ___	14 − 10 = ___
	4 + 10 = ___	14 − 4 = ___
9, ___, and 14	9 + ___ = 14	___ − ___ = ___
	___ + ___ = ___	___ − ___ = ___
8, ___, and 14	8 + ___ = 14	___ − ___ = ___
	___ + ___ = ___	___ − ___ = ___
7, ___, and 14	7 + ___ = 14	___ − ___ = ___
	___ + ___ = ___	___ − ___ = ___

4. Subtract.

a. 13 − 8 = ___	b. 13 − 5 = ___	c. 12 − 7 = ___	d. 12 − 9 = ___
14 − 6 = ___	13 − 4 = ___	13 − 7 = ___	14 − 9 = ___

5. Find the missing numbers.

a. 9 + ☐ = 14	b. 6 + ☐ = 14	c. 6 + ☐ = 12
d. ☐ − 9 = 4	e. ☐ − 7 = 7	f. ☐ − 9 = 3
g. 14 − ☐ = 8	h. 12 − ☐ = 7	i. 13 − ☐ = 8

6. Solve the word problems.

a. Ted arranged his toy cars in rows. The first row had seven cars, the second had seven, and the third row had four. How many cars does Ted have?

b. If you have 14 strawberries and I have eight, how many more do you have?

c. Dad has six cherries and Mom has five more than him. How many cherries does Mom have?

d. At first Mom had 20 apples to make a pie, but she gave each of the four children one apple before she made the pie. How many apples did she have left for the pie?

7. Figure out the patterns and continue them!

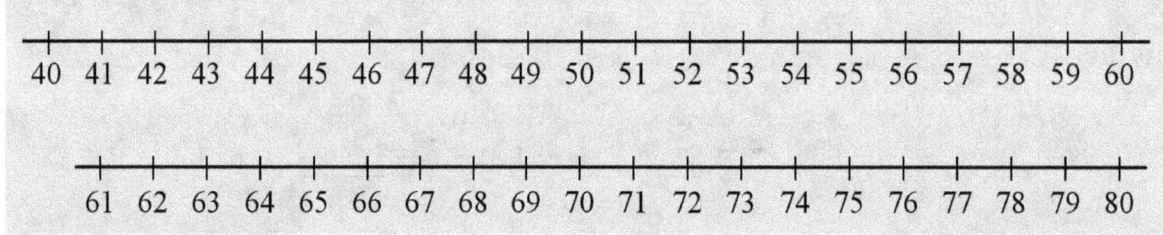

a. 40   48   56   64   72   ___   ___   ___   ___

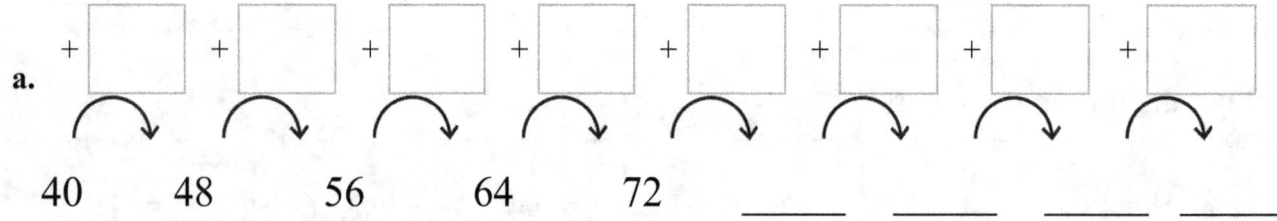

b. 17   21   25   29   ___   ___   ___   ___   ___

Fact Families with 15

1. Fill in. In each fact family, color the marbles so they match the numbers in it.

Fact families with 15		
10, 5, and 15	10 + 5 = ____	15 − 10 = ____
	5 + 10 = ____	15 − 5 = ____
9, ____, and 15	9 + ____ = 15	____ − ____ = ____
	____ + ____ = ____	____ − ____ = ____
8, ____, and 15	8 + ____ = 15	____ − ____ = ____
	____ + ____ = ____	____ − ____ = ____

2. Subtract.

a. 15 − 5 = ____	b. 15 − 8 = ____	c. 15 − 4 = ____
d. 15 − 9 = ____	e. 15 − 6 = ____	f. 15 − 7 = ____

3. Alice does not remember the answer to 15 − 9.
   Explain how she can solve it using *addition*.

4. Count by threes.

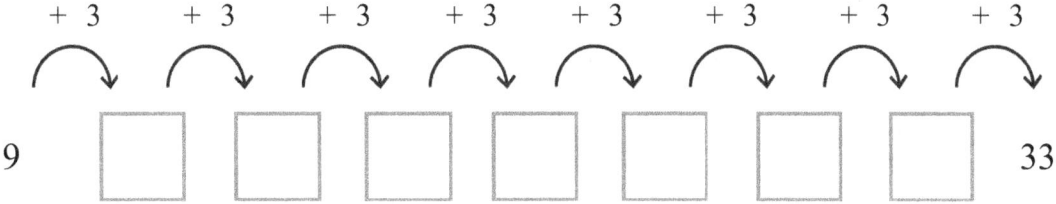

5. These word problems all have to do with "more." Draw a picture of how many things the one person has in the problem. Then think carefully *who* has more. Will you need to draw *more* or *fewer* things for the other person in the problem?

a. Michelle has 7 peaches and Jacob has three more than her. How many does Jacob have?
b. William has three more books than Ethan. William has 11 books. How many does Ethan have?
c. Noah picked 15 pine cones and Aiden picked 9. How many more did Noah pick than Aiden?
d. Emma picked 5 more pine cones than Sophia. If Emma picked 15, how many did Sophia pick?

6. Write each number as a double of some other number.

a. 6 = ____ + ____	**b.** 12 = ____ + ____	**c.** 10 = ____ + ____
d. 18 = ____ + ____	**e.** 20 = ____ + ____	**f.** 8 = ____ + ____

7. Judy picked 7 tomatoes from the garden, and John picked 9. Then they gave half of their tomatoes to a neighbor. How many did they keep?

8. Write or say all the even numbers from 0 to 20.

9. Find how much the things cost together.

a. bike, $28, and kite, $30 together $ _____	**b.** jeans, $47, and shoes, $30, and toy, $10 together $ _____

Fact Families with 16

1. Fill in. Use one color to color the number of marbles for first number, then color the rest another color.

Fact families with 16		
10, 6, and 16	10 + 6 = ___ 6 + 10 = ___	16 − 10 = ___ 16 − 6 = ___
9, ___, and 16	9 + ___ = 16 ___ + ___ = ___	___ − ___ = ___ ___ − ___ = ___
8, ___, and 16	8 + ___ = 16 ___ + ___ = ___	___ − ___ = ___ ___ − ___ = ___

2. Subtract.

a.	b.	c.	d.
15 − 10 = ___	13 − 9 = ___	14 − 8 = ___	15 − 7 = ___
13 − 10 = ___	16 − 9 = ___	13 − 8 = ___	16 − 7 = ___
16 − 10 = ___	14 − 9 = ___	16 − 8 = ___	13 − 7 = ___

3. Connect the problems to the answer with a line.

15 − 9	3	17 − 10	7	17 − 9
14 − 9	4	16 − 9	8	16 − 6
14 − 10	5	16 − 10	9	18 − 10
13 − 9	6	18 − 9	10	19 − 9

4. Figure out the patterns and continue them!

a.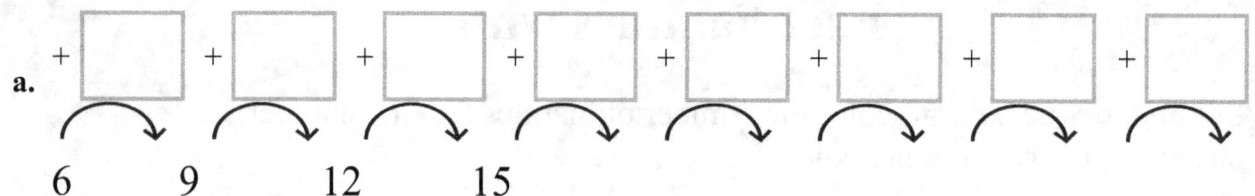
6   9   12   15   ___   ___   ___   ___   ___

b.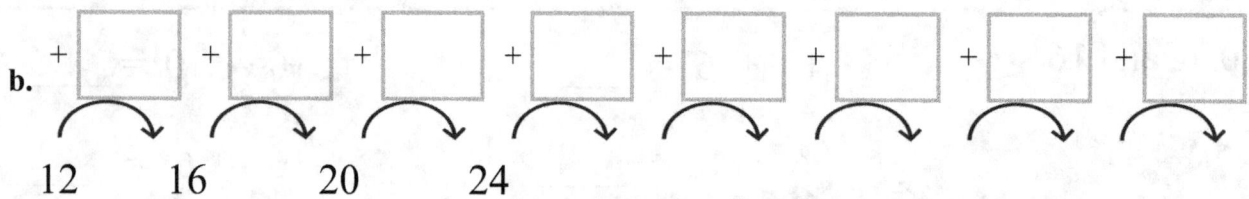
12   16   20   24   ___   ___   ___   ___   ___

5. Solve.

a. A class has 24 children. Two of them were sick one day and two had to leave to go to the dentist. How many children were in class that day?
b. If you have $10, and Mom gives you $4 more, can you buy a book for $13?
c. You had $20 and you bought sandals for $17. How many dollars do you have left?
d. Erika has saved $12. She wants to buy a gift that costs $16. How much more money does she need?
e. Five boys came to play ball. Then, seven girls came. Then, one girl had to go home. Are there now more boys or girls playing ball? How many more?

6. Compare and write <, >, or =.

a. 35 ☐ 20 + 5   b. 23 + 5 ☐ 23 + 6   c. 16 − 8 ☐ 15 − 8

d. 15 ☐ 6 + 7   e. 31 + 4 ☐ 31 + 3   f. 15 − 9 ☐ 16 − 9

Fact Families - 17 and 18

1. Fill in. Use one color to color the number of marbles for first number, then color the rest another color.

Fact families with 17		
10, 7, and 17	10 + 7 = ____	17 − 10 = ____
	___ + ___ = ___	___ − ___ = ___
9, ____, and 17	9 + ____ = 17	___ − ___ = ___
	___ + ___ = ___	___ − ___ = ___

Fact families with 18		
10, 8, and 18	10 + 8 = ____	18 − 10 = ____
	___ + ___ = ___	___ − ___ = ___
9, ____, and 18	9 + ____ = 18	___ − ___ = ___
	___ + ___ = ___	___ − ___ = ___

2. Subtract, practicing the basic facts. Remember to think of fact families.

a.	b.	c.	d.
17 − 10 = ____	15 − 9 = ____	14 − 6 = ____	12 − 9 = ____
17 − 9 = ____	15 − 8 = ____	14 − 7 = ____	12 − 8 = ____
18 − 10 = ____	16 − 9 = ____	13 − 6 = ____	11 − 9 = ____
18 − 9 = ____	16 − 8 = ____	13 − 7 = ____	11 − 8 = ____

3. Write < , > , or = . Can you compare these without calculating?

   a. 45 + 8 ☐ 45 + 5   b. 50 − 6 ☐ 50 − 8   c. $\frac{1}{2}$ of 12 ☐ 12

   d. $\frac{1}{2}$ of 16 ☐ $\frac{1}{2}$ of 14   e. 27 − 6 ☐ 27 − 3   f. $\frac{1}{2}$ of 20 ☐ 10

4. Fill in the missing numbers.

a. 14 − 8 = ◯	b. 16 − 8 = ◯	c. 17 − 8 = ◯
d. ◯ − 9 = 6	e. ◯ − 8 = 7	f. ◯ − 4 = 8
g. 17 − ◯ = 9	h. 18 − ◯ = 9	i. 15 − ◯ = 6

5. Solve the word problems.

a. A baby slept four hours and woke up to nurse. Then she slept another two hours and woke up to nurse. Then she slept three hours more and nursed again. Then she slept three hours until the morning.

How many hours did the baby sleep?

b. Mom needs 16 eggs to make cakes. The store sells eggs in cartons of 12. How many cartons does she need to buy?

How many eggs will she have left over?

6. Find the missing numbers. You can also work backwards, starting from 70!

 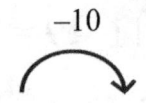 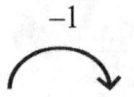 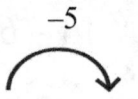 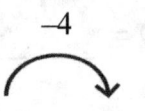 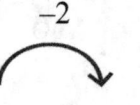 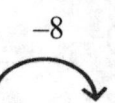 

100 ___ ___ ___ ___ ___ 70

Mixed Review Chapter 3

1. Write the time.

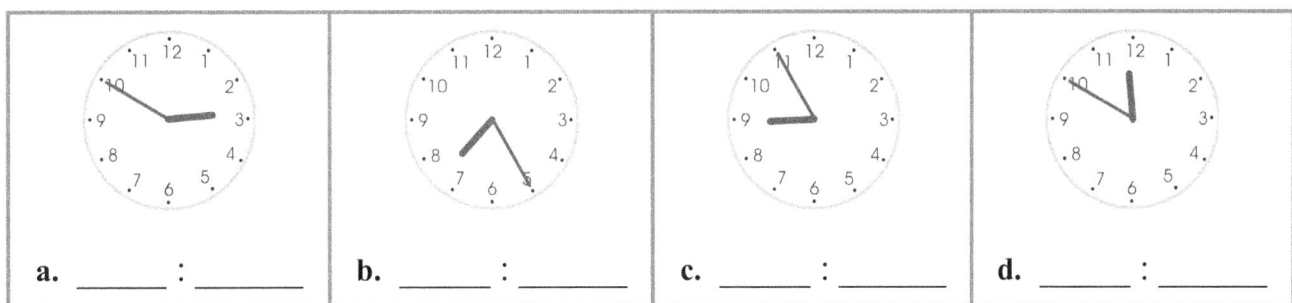

a. ____ : ____   b. ____ : ____   c. ____ : ____   d. ____ : ____

2. Write the time that the clock shows, and the time 5 minutes later.

	a. ____ : ____	b. ____ : ____	c. ____ : ____	d. ____ : ____
5 min. later →	____ : ____	____ : ____	____ : ____	____ : ____

3. How many minutes pass? Subtract (or figure out the difference).

from	2:25	2:20	7:00	11:30	6:05
to	2:35	2:40	7:15	11:50	6:15
minutes	10 minutes				

4. Ashley shared 18 raisins and 12 almonds equally with her brother.
   How many raisins did each child get?
   How many almonds?

5. Write each number as a double of some other number.

a. 10 = ____ + ____   b. 16 = ____ + ____   c. 40 = ____ + ____

6. Fill in the missing numbers for this subtraction "journey".

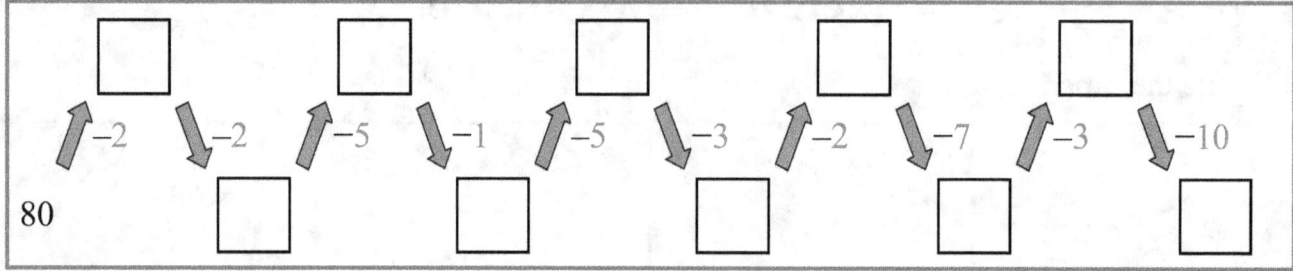

80

7. Solve the problems.

a. Mary ate 20 strawberries and Isabella ate half that many.

   How many did Isabella eat?

   How many did the girls eat together?

b. Kyle used half of his money to buy a toy car. Now he has $10 left.
   How much money did Kyle have at first?

c. Jane ate 10 strawberries more than what Jonathan ate.
   Jonathan ate 12 strawberries. How many did Jane eat?

d. Emily is 30 years old, and Hannah is 4 years old.
   How many years older is Emily than Hannah?

e. Ann had 12 toy cars, and Judith had 10. Then Ann got two more cars.

   Now who has more cars?

   How many more?

f. Jacob has $6 and Jim has $7 more than Jacob.
   How much money does Jim have?

Review Chapter 3

1. Here are the 20 addition facts with single-digit numbers where the sum is between 10 and 20. Connect the problems to the right answer.

   5 + 6   4 + 8   6 + 9
   6 + 8   11   6 + 7   15   8 + 8
   6 + 6   12   9 + 4   16   7 + 8
   4 + 7   7 + 7   9 + 8
           13   17
   3 + 9   2 + 9   7 + 9
   3 + 8   14   5 + 7   18
   8 + 5   5 + 9   9 + 9

2. Connect with a line the problems that are from the same fact family. You don't need to write the answers.

13 − 7 = ☐	12 − 5 = ☐	15 − 7 = ☐
7 + ☐ = 15	11 − 8 = ☐	13 − 6 = ☐
11 − 3 = ☐	9 + ☐ = 17	5 + ☐ = 14
8 + ☐ = 17	15 − 8 = ☐	17 − 8 = ☐
14 − 5 = ☐	6 + ☐ = 13	3 + ☐ = 11
7 + ☐ = 12	9 + ☐ = 14	☐ + 5 = 12

3. Find the differences.

a. The difference of 80 and 87 _____	**b.** The difference of 45 and 2 _____
c. The difference of 15 and 8 _____	**d.** The difference of 13 and 4 _____

4. Find the missing numbers.

a. $8 + \square = 15$	b. $7 + \square = 14$	c. $6 + \square = 13$
d. $13 - \square = 5$	e. $14 - \square = 8$	f. $15 - \square = 9$
g. $11 - 6 = \square$	h. $12 - 7 = \square$	i. $12 - 4 = \square$

5. Find the missing steps.

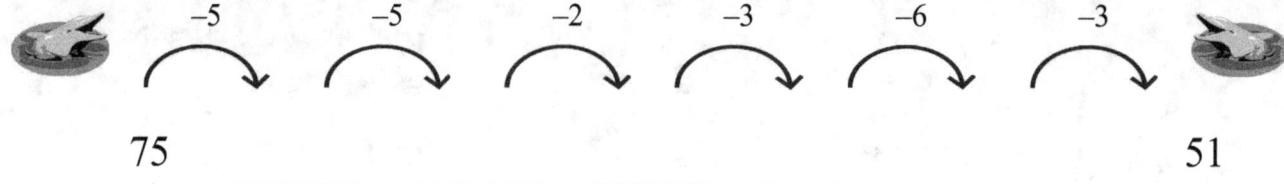

75 _____ _____ _____ _____ _____ 51

6. a. You have an *odd* number of cookies and so does your friend. You put your cookies together and share them. Can you share them evenly or not?

Cookies you have	Cookies your friend has	Together we have	even/odd	Can you share evenly?
3	5			
5	9			
9	3			
9	7			

b. You have an *odd* number of cookies and your friend has an *even* number of cookies. You put your cookies together and share them. Can you share them evenly or not?

Cookies you have	Cookies your friend has	Together we have	even/odd	Can you share evenly?
5	6			
7	8			
9	4			
1	12			

7. Solve the puzzle. What happened to the teddy bear in the desert?

5 + 9   7 + 8   13 − 8   2 + 9   10 + 5   9 + 7   4 + 7   9 + 6

___ ___   ___ ___ ___   ___ ___ ___

7 + 7   13 − 6   19 − 4   11 + 5   13 − 7   3 + 13   11 − 5   13 − 4   6 + 9

___ ___   ___ ___ ___   ___ ___ ___ ___

Key:

A	E	I	O	G	H	T	W	N
9	6	14	11	5	16	15	8	7

8. Solve the word problems.

a. Jack has 13 tennis balls and Jane has 20.
How many more does Jane have than Jack?

b. Emma has three more flowers than Sofia. If Emma has 14 flowers, how many does Sofia have?

c. In a chess game, Jacob has 2 more pawns than Anna. If Anna has five pawns, how many does Jacob have?

d. You have $20, and you want to buy a Lego set that costs $28. How many dollars do you still need to save?

Later, a neighbor pays you $2 for helping rake leaves. How much more money do you need after that?

e. In a board game, you need to move 18 more squares to get to the end of the game. You roll 6 and 5 on two dice and move that many squares. Now how many more squares are there to the end?

What kind of numbers on the two dice would get you to the end?

Chapter 4: Regrouping in Addition
Introduction

The fourth chapter of *Math Mammoth Grade 2* deals with addition within 0-100, both mentally and in columns, especially concentrating on regrouping in addition (carrying).

Mental math is important because it builds number sense, and in turn, number sense develops algebraic thinking. We study how to add mentally a two-digit number and a single-digit number (for example 36 + 8 or 45 + 9). To do that, children learn to use a "helping problem" composed of the single-digit numbers (6 + 8 or 5 + 9). Just like 6 + 8 fills the ten and is four more than that, even so, the sum 36 + 8 fills the *next* whole ten (40), and is four more than that, or 44.

We also study adding two-digit numbers with regrouping (aka "carrying"). This process is illustrated and explained in detail with the help of visual models. You are welcome to also use actual manipulatives if you prefer. The main concept here is that 10 ones make a new ten. This new ten is regrouped with the other tens, and written using a little "1" in the tens column.

In order to prepare for adding three or four two-digit numbers in columns, we practice explicitly how to add three or four single-digit numbers, such as 7 + 8 + 6 + 4, and the principle of adding in parts (such as 13 + 16 is the same as 10 + 10 and 3 + 6).

The lessons also include lots of word problems, a review of even and odd numbers, and occasional review problems about doubling. Once again, don't automatically assign all the problems and exercises, but use your judgment.

Pacing Suggestion for Chapter 4

Please add one day to the pacing for the test if you use it. Note that the specific lessons in the chapter can take several days to finish. They are not "daily lessons." As a general guideline, second graders should finish 1.5 to 2 pages daily or 8 to 10 pages a week. Please also see the user guide at
https://www.mathmammoth.com/userguides/ .

The Lessons in Chapter 4	page	span	suggested pacing	your pacing
Going Over to the Next Ten	101	*3 pages*	2 days	
Add with Two-Digit Numbers Ending in 9	104	*2 pages*	1 day	
Add a Two-Digit Number and a Single-Digit Number Mentally	106	*2 pages*	1 day	
Regrouping with Tens	108	*3 pages*	2 days	
Add in Columns Practice	111	*3 pages*	2 days	
Mental Addition of Two-Digit Numbers	114	*3 pages*	2 days	
Adding Three or Four Numbers Mentally	117	*2 pages*	1 day	
Adding Three or Four Numbers in Columns	119	*4 pages*	2 days	
Mixed Review Chapter 4	123	*2 pages*	1 day	
Review Chapter 4	125	*2 pages*	1 day	
Chapter 4 Test (optional)				
TOTALS		*26 pages*	15 days	

Games and Activities

Make 100

You need: Two standard decks of playing cards from which you remove the face cards and tens, leaving only numbers 1 through 9.

Game Play: In each round, each player is dealt four cards. Each player forms two 2-digit numbers with his four cards, using each card as a digit. For example, if you're dealt 4, 8, 6, and 1, you could make 84 and 16. Or, you could make 41 and 68. The goal is to make these two numbers in such a manner that their sum is as close to 100 as possible. Each player calculates the sum of their numbers <u>mentally</u>. The player with the sum closest to 100 wins that round, and puts all the cards played on that round to his personal pile.

In the case of a tie, the players are dealt four new cards each, and they use those to resolve the tie. After enough rounds have been played to use all of the cards in the deck, the player with the most cards in his personal pile wins.

Variation: Allow players to calculate the sums using pencil and paper. Mental math is much faster, though. (You can always add the tens separately and the ones separately, and add those two sums.)

Simple Dice

You need: five six-sided dice.

The goal of the game to is to get the maximum sum from the five dice. The game practices mental addition of several small numbers.

Game play: At your turn, roll the five dice. You have to leave at least one of the dice (hold it), but you may reroll up to four of them. Again, you have to hold at least one dice, and you can reroll the rest. After these three rolls, your turn is over. Calculate the sum of your dice. This is then written down as your score for this turn.

After a set number of turns (such as five), each player calculates their total score of all the rounds. The player with the highest total wins.

One Is IN

This is a variation of the above game, Simple Dice. It adds in one additional rule, and that is why I recommend that you first play the Simple Dice game with your child or students, so they learn the basic idea of the game.

You need: five 6-sided dice

The goal of the game to is to get the maximum sum from the five dice. One of the dice has to show 1, for you to score at all.

Game play: At your turn, roll the five dice. You have to leave at least one of the dice (hold it), but you may reroll up to four of them. Again, you have to hold at least one dice, and you can reroll the rest. After four such rolls, your turn is over. If at least one of your dice shows 1, calculate the sum of your dice. This is then written down as your score for this turn. If none of your dice show 1, you do not score anything.

After a set number of turns (such as five), each player calculates their total score of all the rounds. The player with the highest total wins.

Games and Activities at Math Mammoth Practice Zone

Two-Digit Addition with Mental Math
Simple online practice of adding two-digit numbers using mental math.

- Add a two-digit and a single-digit number:
  https://www.mathmammoth.com/practice/addition-subtraction-two-digit#opts=2p1dwr
- Add two 2-digit numbers, no regrouping:
  https://www.mathmammoth.com/practice/addition-subtraction-two-digit#opts=2p2dnr
- Add two 2-digit numbers, with regrouping:
  https://www.mathmammoth.com/practice/addition-subtraction-two-digit#opts=2p2dwr

Hidden Picture Addition Game
Add two-digit numbers and reveal a hidden picture.
https://www.mathmammoth.com/practice/mystery-picture#min=11&max=99

Mathy's Berry Picking Adventure
The first link practices adding a two-digit and a single-digit number (e.g. 45 + 7). The second link practices mentally adding two 2-digit numbers (e.g. 34 + 26).

- https://www.mathmammoth.com/practice/mathy-berries#mode=addition-both&duration=2m
- https://www.mathmammoth.com/practice/mathy-berries#mode=addition-double&duration=2m

Bingo
For this chapter, choose Addition (Two-Digit) to practice mental addition of two-digit numbers.
https://www.mathmammoth.com/practice/bingo

Fruity Math
Click the fruit with the correct answer and try to get as many points as you can within two minutes.

- Add a two-digit number and nine:
  https://www.mathmammoth.com/practice/fruity-math#op=addition&duration=120&mode=manual&config=12,89x1__9,9x1&max-sum=200
- Add a two-digit and a single-digit number:
  https://www.mathmammoth.com/practice/fruity-math#op=addition&duration=120&mode=manual&config=13,89x1__3,9x1&max-sum=200
- Add two 2-digit numbers:
  https://www.mathmammoth.com/practice/fruity-math#op=addition&duration=120&mode=manual&config=11,89x1__11,99x1&max-sum=125
- Start with single-digit additions, and then advance through levels with increasingly harder sums:
  https://www.mathmammoth.com/practice/fruity-math#op=addition&duration=120&mode=levels&start-level=2

Make number sentences
Drag two flowers to the empty slots to make the given sum, practicing two-digit mental addition.
https://www.mathmammoth.com/practice/number-sentences#questions=5&types=add-11-80

Color-Grid Game — Vertical Addition Practice
Solve 12 problems of adding two-digit numbers in columns.
https://www.mathmammoth.com/practice/vertical-addition#max=99&questions=4*3&addends=2&max-digits=3

Further Resources on the Internet

These resources match the topics in this chapter, and offer online practice, online games (occasionally, printable games), and interactive illustrations of math concepts. We heartily recommend you take a look. Many people love using these resources to supplement the bookwork, to illustrate a concept better, and for some fun. Enjoy!

https://l.mathmammoth.com/gr2ch4

Scan me

Going Over to the Next Ten

Sums that go over to the next ten

Let's add 59 + 5 so that we *first* complete 60.

59 + 5
    | \
59 + 1 + 4

  60 + 4 = 64

The 5 is broken into two parts: 1 and 4.
That is because 59 and 1 makes sixty.
Then, we have 60 and 4. We get 64.

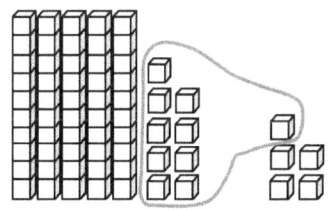

9 and 1 make a ten.
We get 6 tens.

59 + 5 = 64

1. Circle ten little cubes to make a ten. Count the tens and ones. Write the answer.

a. 13 + 9 = _____

b. 15 + 8 = _____

c. 17 + 7 = _____

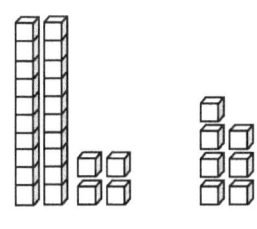

d. 24 + 7 = _____

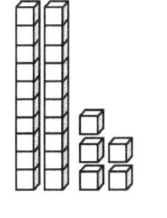

e. 25 + 6 = _____

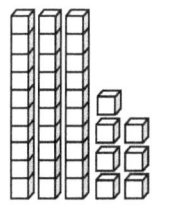

f. 37 + 9 = _____

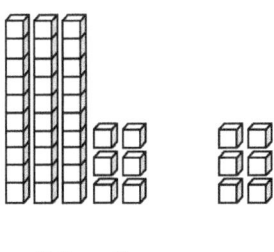

g. 36 + 6 = _____

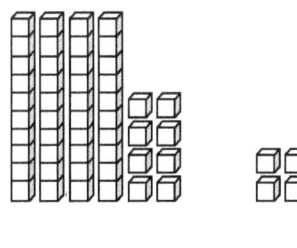

h. 48 + 4 = _____

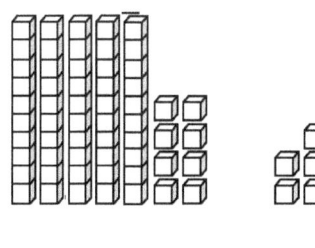

i. 58 + 5 = _____

2. Complete. Break the second number into two parts so that you complete the next ten.

a. 28 + 8
   / \
28 + _2_ + ___
 	
30 + ___ = ___
	b. 47 + 5
	   / \
	47 + _3_ + ___
	
	50 + ___ = ___
		c. 79 + 9
		   / \
		79 + ___ + ___
		
		80 + ___ = ___

d. 39 + 3
   / \
39 + ___ + ___
 	
40 + ___ = ___
	e. 27 + 5
	   / \
	27 + ___ + ___
	
	___ + ___ = ___
		f. 38 + 7
		   / \
		38 + ___ + ___
		
		___ + ___ = ___

3. Write the additions illustrated by the number-lines. Think how long each line is.

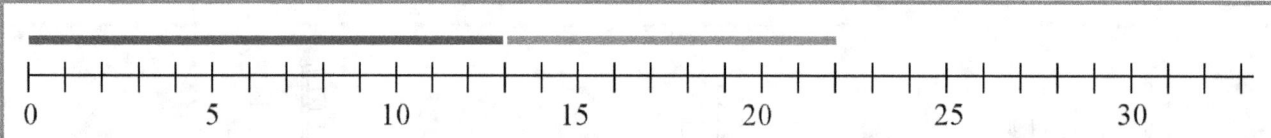

a. ___ + ___ = ___

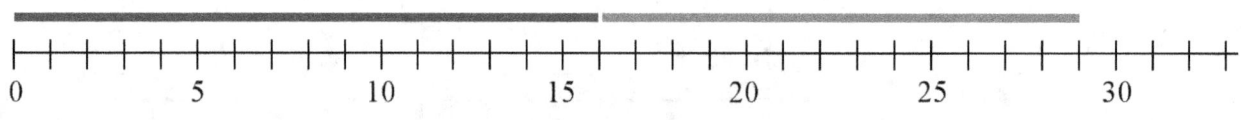

b. ___ + ___ = ___

4. Show these additions on the number line by drawing two lines.

a. 19 + 7 = ___

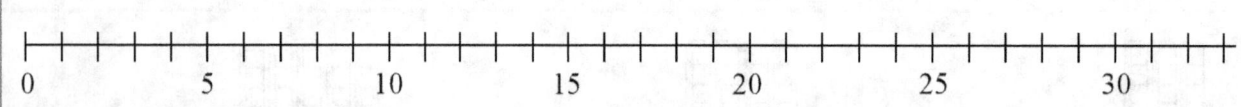

b. 14 + 18 = ___

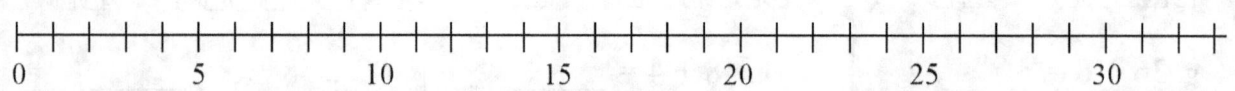

5. Solve the problems. Write a number sentence for each problem, not just the answer.

   a. Benjamin wants to buy a kite that costs $30. He has saved $22. How much more money will he need?

   b. Natasha had already saved $20. She earned $5 by selling eggs, and earned $5 more for selling fruit. How much money does she have now?

   c. Mom bought 28 fruit trees and has planted eight of them. How many still need planted?

   d. Thirty-seven people attended Uncle Sam's 50th birthday party. Thirty-two of them came before noon. How many came after noon?

   e. Dad bought a bunch of 40 grapes and ate half of them. Then, little sister ate seven grapes. How many are left now?

6. Continue the patterns. COMPARE the columns, and NOTICE what is the same.

a. 8 + 1 = ____	b. 28 + 1 = ____	c. 78 + 1 = ____
8 + 2 = ____	28 + 2 = ____	78 + 2 = ____
8 + 3 = ____	28 + 3 = ____	78 + 3 = ____
8 + 4 = ____	28 + 4 = ____	78 + 4 = ____
8 + ____ = ____	28 + ____ = ____	78 + ____ = ____
8 + ____ = ____	28 + ____ = ____	78 + ____ = ____

Add with Two-Digit Numbers Ending in 9

Imagine that 29 wants to be 30...
so it "grabs" one from 5.
Then, 29 becomes 30, and 5 becomes 4.

The addition problem is changed to 30 + 4 = 34.

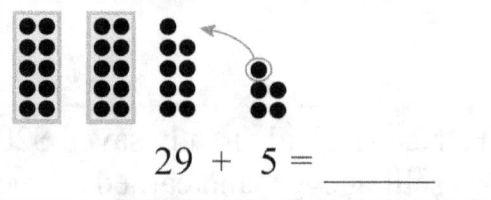

29 + 5 = _____

1. Circle the nine dots and one more dot to form a complete ten. Add.

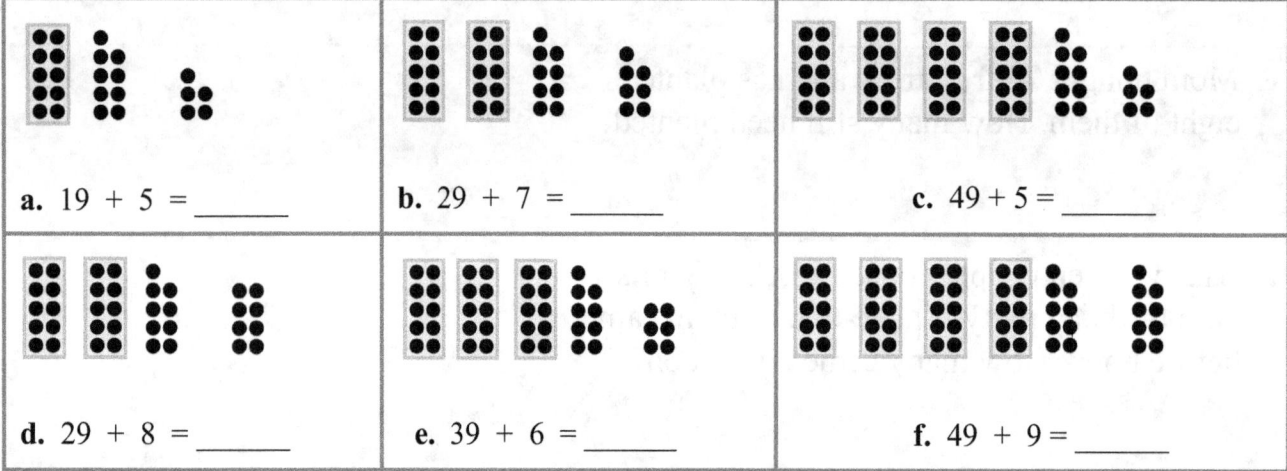

a. 19 + 5 = _____   b. 29 + 7 = _____   c. 49 + 5 = _____

d. 29 + 8 = _____   e. 39 + 6 = _____   f. 49 + 9 = _____

2. Add. For each problem, write a helping problem using the "ones" from the first problem.

a. 19 + 7 = _____   b. 49 + 3 = _____   c. 39 + 4 = _____

  9 + 7 = _____   ___ + ___ = ___   ___ + ___ = ___

3. Add. Compare the problems.

a. 9 + 3 = _____   b. 9 + 6 = _____   c. 9 + 4 = _____

   19 + 3 = _____   39 + 6 = _____   49 + 4 = _____

d. 9 + 7 = _____   e. 9 + 9 = _____   f. 9 + 5 = _____

   39 + 7 = _____   69 + 9 = _____   19 + 5 = _____

   29 + 7 = _____   79 + 9 = _____   59 + 5 = _____

4. These problems review the basic facts with 9 and 8. By this time you should already remember these addition facts. Try to remember what number will fit without counting.

a. 	b. 	c. 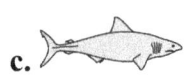	d.
14 − 9 = ___	4 + 9 = ___	15 − ___ = 8	7 + 8 = ___
15 − 9 = ___	8 + 9 = ___	17 − ___ = 8	5 + 8 = ___
13 − 9 = ___	5 + 9 = ___	12 − ___ = 8	6 + 8 = ___
18 − 9 = ___	6 + 9 = ___	14 − ___ = 8	3 + 8 = ___
17 − 9 = ___	9 + 9 = ___	13 − ___ = 8	9 + 8 = ___
16 − 9 = ___	7 + 9 = ___	16 − ___ = 8	4 + 8 = ___

5. Find the difference of numbers. The number line can help.

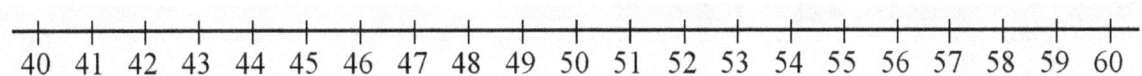

a. Difference between 41 and 53 ___	b. Difference between 60 and 46 ___	c. Difference between 59 and 48 ___

6. Find the patterns and continue them!

a. 0   1   3   6   10   ___   ___   ___   ___

b. ___   ___   ___   ___   ___   44   48   52   56

Add a Two-Digit Number and a Single-Digit Number Mentally

Imagine that 38 wants to be 40, so it "grabs" two from 7. Then, 38 becomes 40, and 7 becomes 5.

The addition problem is changed to 40 + 5 = 45.

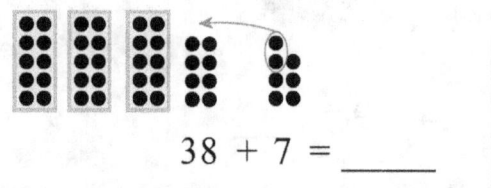

38 + 7 = _____

1. Circle the eight dots and two more dots to form a complete ten. Add.

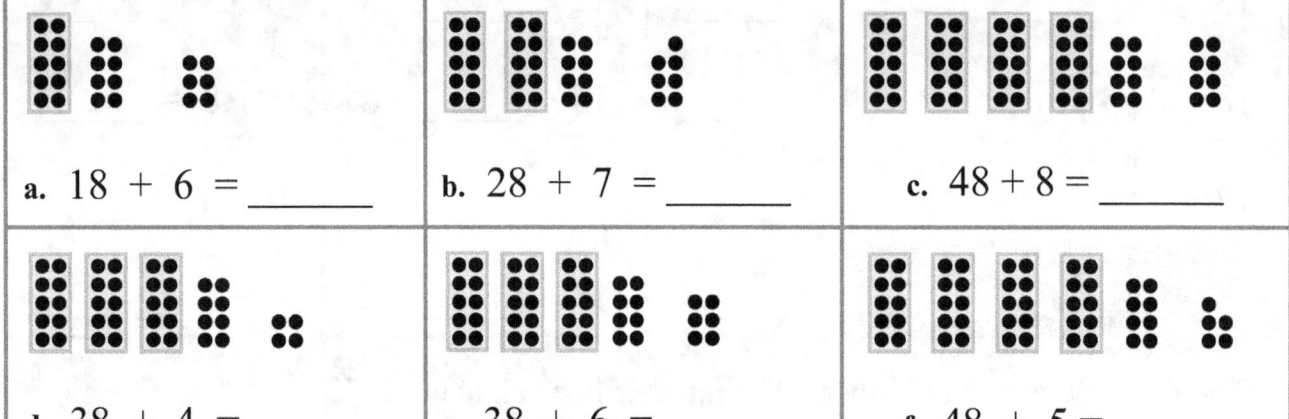

a. 18 + 6 = _____   b. 28 + 7 = _____   c. 48 + 8 = _____

d. 38 + 4 = _____   e. 38 + 6 = _____   f. 48 + 5 = _____

2. Add. Think of the trick explained above.

a. 18 + 7 = _____   b. 38 + 6 = _____   c. 58 + 5 = _____

3. Add. Compare the problems. What is similar about the problems in each box?

a. 8 + 3 = _____	b. 8 + 6 = _____	c. 8 + 4 = _____
18 + 3 = _____	38 + 6 = _____	78 + 4 = _____
d. 8 + 2 = _____	e. 8 + 9 = _____	f. 8 + 5 = _____
38 + 2 = _____	68 + 9 = _____	18 + 5 = _____
28 + 2 = _____	78 + 9 = _____	58 + 5 = _____

When you add a two-digit number and a single-digit number, such as 45 + 6 or 77 + 4, think of the "helping" problem: the addition with just the ones digits.

Example. 45 + 6	**Example. 67 + 8**
Think of the helping problem 5 + 6 = 11. (Drop the 40 from 45, and you have 5 + 6.) 5 + 6 is ONE more than the next ten (11), and 45 + 6 is also ONE more than the next ten (51).	Think of the helping problem 7 + 8 = 15. (Drop the 60 from 67, and you have 7 + 8.) 7 + 8 is FIVE more than the next ten (15), and 67 + 8 is also FIVE more than the next ten (75).

4. Add and compare! The top problem is a helping problem for the bottom one.

a. 7 + 6 = _____	b. 6 + 8 = _____	c. 7 + 7 = _____
27 + 6 = _____	76 + 8 = _____	87 + 7 = _____
(three more than the next ten)	(four more than the next ten)	(four more than the next ten)
d. 5 + 8 = _____	e. 6 + 9 = _____	f. 8 + 7 = _____
35 + 8 = _____	26 + 9 = _____	48 + 7 = _____

5. Fill in: To add 73 + 8, I can use the helping problem ___ + ___ = _____. Since the answer to that is ___ more than 10, the answer to 73 + 8 is ___ more than ___.

6. Add.

a. 34 + 8 = _____	b. 47 + 7 = _____	c. 59 + 5 = _____

7. Solve the word problems.

a. Jenny needed 24 eggs to make omelets for her family. She already had 10 eggs. How many more does she need?
b. Her large family eats lots of potatoes. Dad bought a 25-kilogram bag of potatoes. Now, only 5 kg are left. How many kilograms of potatoes have they eaten?

Regrouping with Tens

When adding 3 + 9, we can circle ten little ones to form a ten. We write "1" in the tens column.

There are two little ones left over, so we write "2" in the ones column.

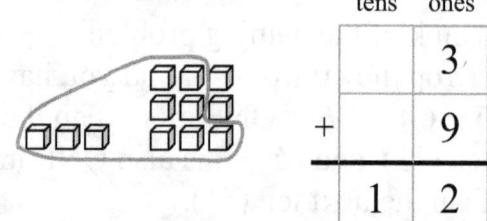

tens	ones
	3
+	9
1	2

With 35 + 8, we circle ten little ones to make a ten. There already are three tens, so in total we now have <u>four</u> tens. So, we write "4" in the tens column.

There are three little cubes left over, so we write "3" in the ones column.

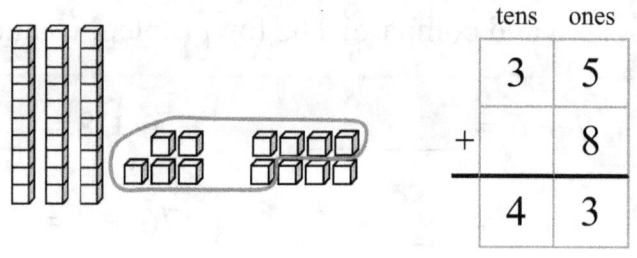

tens	ones
3	5
+	8
4	3

1. **Circle** ten cubes to make **a new ten**. Count the tens, including the new one. Count the ones. Write the tens and ones in their own columns. You can use manipulatives.

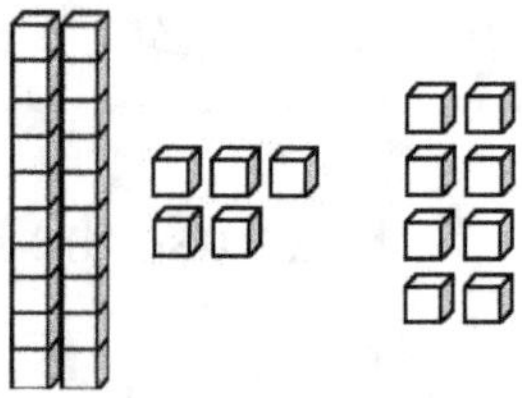

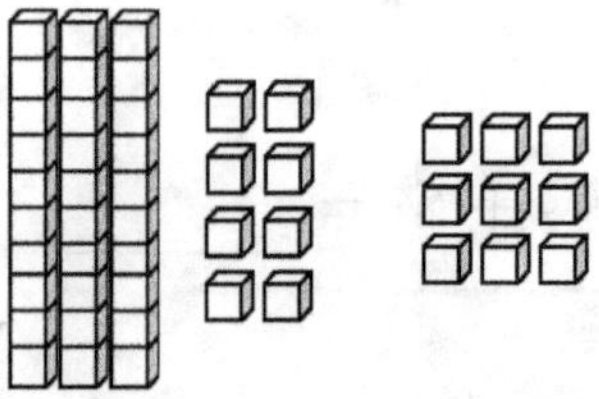

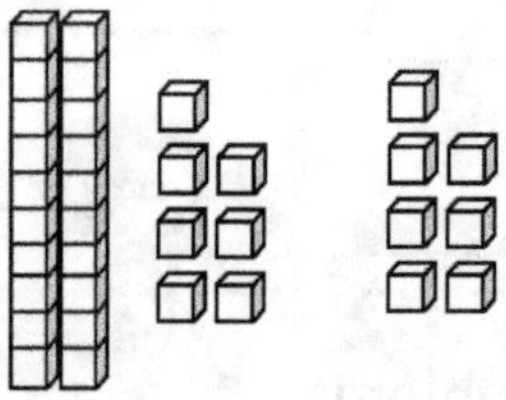

a. 3 3 + 9

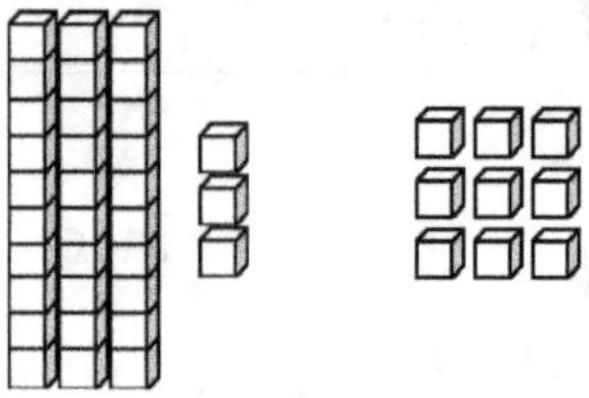

b. 2 5 + 8

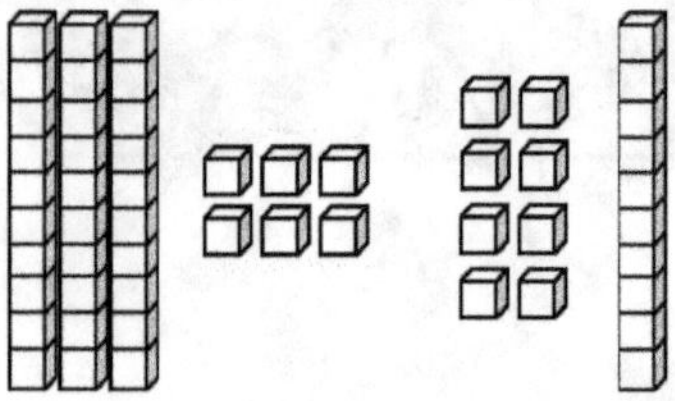

c. 3 8 + 9

d. 2 7 + 7

e. 3 6 + 1 8

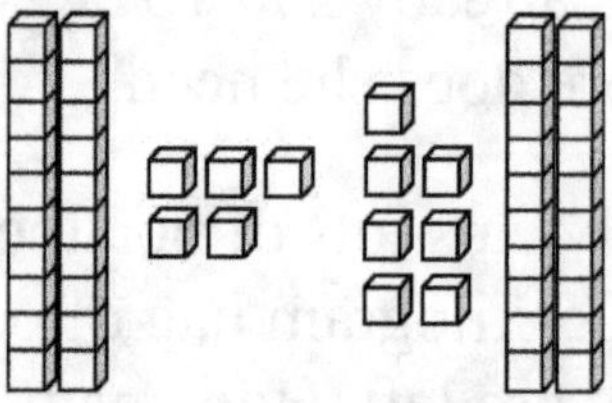

f. 2 5 + 2 7

108

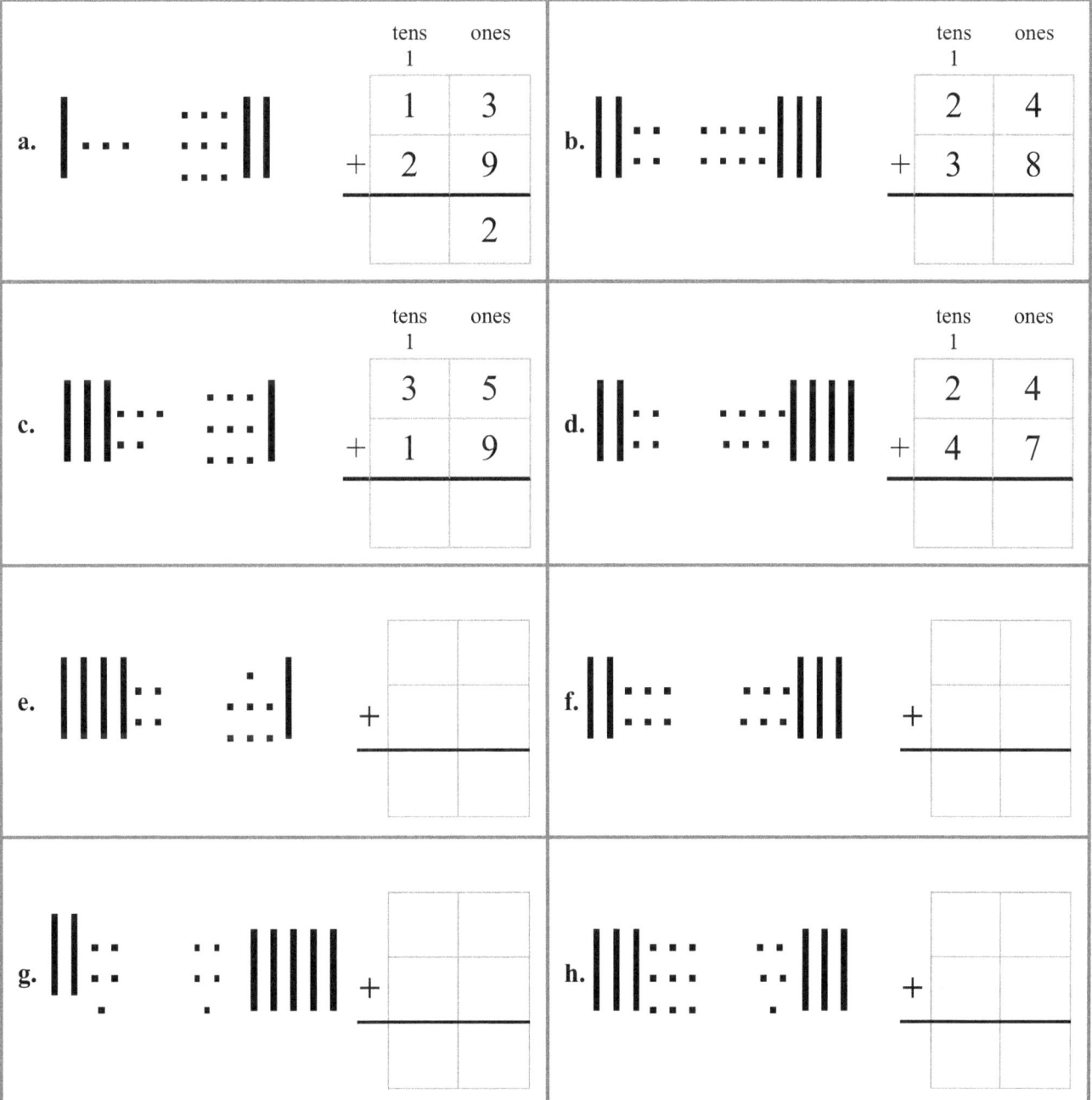

3. Add. If you can make a new ten from the ones, regroup.

a. 42
   +15
   ———

b. 27
   +45
   ———

c. 65
   +26
   ———

d. 83
   +15
   ———

e. 34
   +19
   ———

f. 52
   +41
   ———

g. 13
   +44
   ———

h. 63
   +27
   ———

i. 36
   +51
   ———

j. 66
   +29
   ———

We can add three numbers by writing them under each other. This is not any more difficult than adding two numbers.

On the right, first add the ones. 2 + 7 + 5 = 14. You get a new ten. So, regroup and write that new ten with the other tens.

In the tens, add 1 + 3 + 2 + 1 = 7.

	1	
	3	2
	2	7
+	1	5
	7	4

---

4. Add. Regroup the ones to make a new ten.

a. 34
    19
   +26
   ———

b. 15
    27
   +45
   ———

c. 13
    27
   +26
   ———

d. 26
    42
   +19
   ———

e. 34
    21
   +19
   ———

5. Show the additions on the number line by drawing lines that are that long.

a. 13 + 9 + 11 = _____

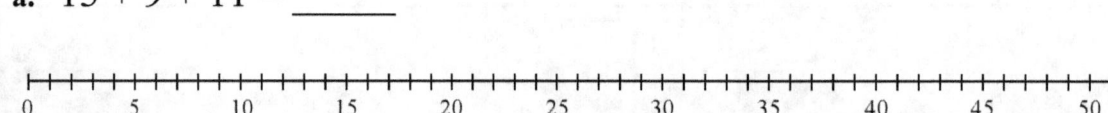

b. 27 + 16 = _____

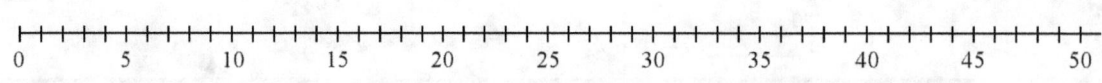

Add in Columns Practice

1. Add in columns.

a.	b.	c.	d.	e.
9 + 7 1	2 4 + 6 7	5 5 + 3 6	4 5 + 2 5	3 8 + 1 4

f.	g.	h.	i.	j.
3 4 9 + 3 5	2 5 4 2 + 4 9	5 8 3 0 + 6	2 9 4 4 + 1 2	1 6 1 4 + 1 9

2. Write the numbers so that the ones and tens are in their own columns. Add.

a. 45 + 27   **b.** 8 + 56   **c.** 40 + 32   **d.** Double 35   **e.** Double 47

tens ones   tens ones

f. 6 + 31 + 25   **g.** 40 + 7 + 9   **h.** 46 + 8 + 20   **i.** 5 + 8 + 13   **j.** 5 + 4 + 57

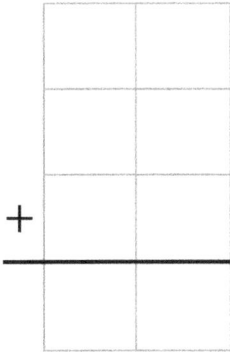
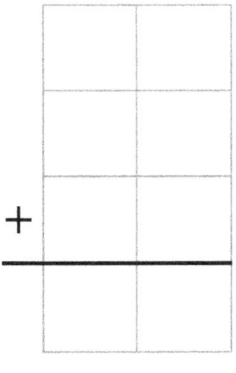
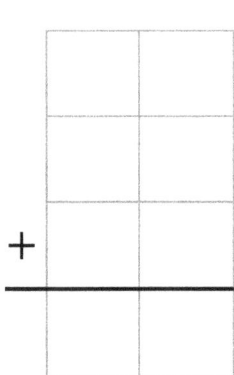
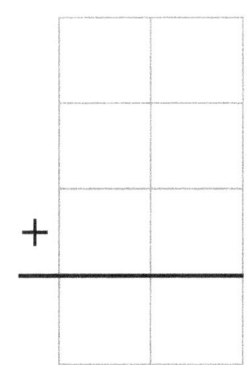
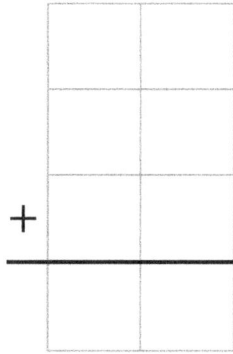

Here we have more than 10 tens.

Ten tens make a *hundred* (100)!

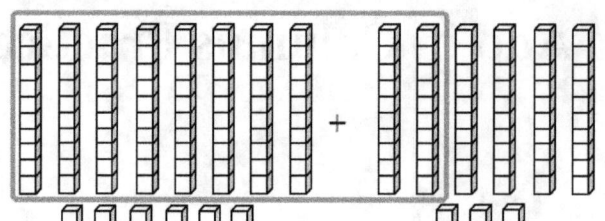

	hund-reds	tens	ones
		8	6
+		6	3
	1	4	9

Add the tens: 8 + 6 = 14 tens. The "1" of the 14 goes in the hundreds column, and the "4" stays in the tens column. The answer 149 is read "one hundred and forty-nine."

Another example. Add the tens normally: 1 + 5 + 6 = 12 tens. Write the 12 so that the "1" is in the hundreds' column, and the "2" is in the tens column. The 12 tens make 1 hundred and 2 tens.

The answer 123 is read "one hundred and twenty-three."

You will study more about hundreds later.

	hund-reds	tens	ones
		1	
		5	4
+		6	9
	1	2	3

3. Add. You will have more than 10 tens.

    a. 27 + 80   **b.** 95 + 47   **c.** Double 56   **d.** 62 + 84

4. Add.

a. 67 + 61	**b.** 90 + 65	**c.** 39 + 81	**d.** 85 + 62	**e.** 29 + 94

f.   65
     18
+ 26
	g. 74
	      7
	+ 45
		h. 68
		     47
		+ 32
			i. 12
			     88
			+ 49
				j. 8
				    50
				+ 79

5. Solve the word problems. You may need to add or subtract in columns.

a. Josh worked for 27 hours this week.
Bill worked for 16 hours more than Josh.
How many hours did Bill work?

b. Natasha read 29 comic books and Matt read 16.
How many more comic books did Natasha read than Matt?

c. Mom put 13 red flowers and 11 blue flowers in one vase. Then she put 22 flowers in another vase.
Which vase has more flowers?

How many more?

d. Caleb had saved $24 and his brother David $41.
Then Caleb earned $20.
Now, who has more money?

How much more?

e. Caleb bought a set of colored pencils for $13, drawing paper for $9, and paints for $21.
What was the total cost?

Mental Addition of Two-Digit Numbers

Example 1. Add in parts 40 + 55.

First break 55 into its tens and ones. 55 is 50 + 5.

So, 40 + 55 becomes 40 + 50 + 5.

Now add 40 and 50. You get 90. Then add the 5. You get 90 + 5 = _____.

Example 2. Add in parts 36 + 30.

First break 36 into three tens and ones. 36 is 30 + 6.

So, 36 + 30 becomes 30 + 6 + 30.

Now add 30 and 30. That is 60. Then add the 6. You get 60 + 6 = _____.

1. Add *in parts*, breaking the second number into its tens and ones.

a. 20 + 34 = _____	b. 70 + 18 = _____	c. 50 + 27 = _____
20 + ____ + ____	70 + ____ + ____	50 + ____ + ____

2. Add *in parts*. Break the number that is not whole tens into its tens and ones in your mind.

a. 17 + 10 = _____	b. 16 + 20 = _____	c. 50 + 14 = _____
26 + 10 = _____	34 + 30 = _____	60 + 23 = _____
42 + 10 = _____	67 + 20 = _____	30 + 45 = _____

3. Add mentally. We already studied these. The first one is the helping problem.

a.	b.	c.	d.
7 + 8 = _____	4 + 9 = _____	8 + 4 = _____	7 + 9 = _____
17 + 8 = _____	14 + 9 = _____	48 + 4 = _____	57 + 9 = _____
37 + 8 = _____	44 + 9 = _____	78 + 4 = _____	37 + 9 = _____

| **How can you easily add 16 + 19?** Think about it before you go on! Here is the answer: again, add *in parts*. Look at the example on the right. | 16 + 19
= 6 + 9 + 10 + 10
= 15 + 10 + 10 = ____ |

4. Add in parts.

a. 13 + 18 = ___ + ___ + 10 + 10 =	b. 15 + 15 = ___ + ___ + 10 + 10 =
c. 17 + 18 = ___ + ___ + 10 + 10 =	d. 19 + 15 = ___ + ___ + 10 + 10 =
e. 18 + 12 = ____	f. 13 + 16 = ____
g. 16 + 17 = ____	h. 17 + 15 = ____

5. **a.** Laura owns 13 cats. Five of her cats live in the house. How many of her cats live outside?

   b. Laura's cats eat 20 lb of cat food in a week. Laura has *two* 4-lb bags at home. How many more pounds of cat food does she need to have enough for one week?

6. Count by threes.

   42, 45, _____, _____, _____, _____, _____, _____, _____

7. Find the pattern and continue it. This pattern "grows" at each step.

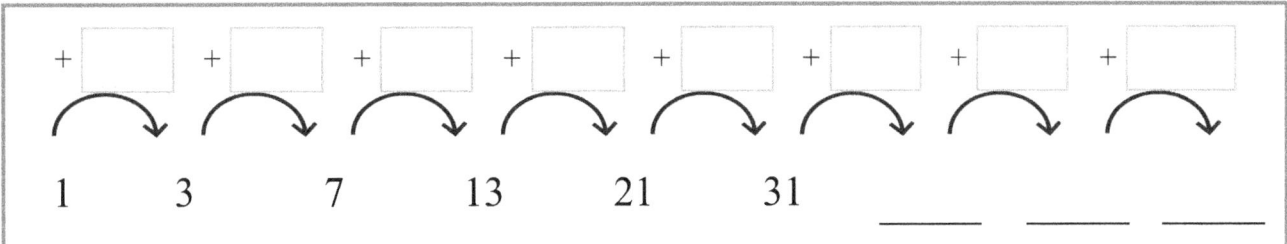

Add two-digit numbers: Add the tens and ones separately	
Add tens on their own. Add ones on their own. Lastly, add the two sums.	45 + 27 40 + 20 + 5 + 7 60 + 12 = 72

8. Add by adding tens and ones separately.

a. 36 + 22 30 + 20 + 6 + 2 ____ + ____ = ____	**b.** 72 + 18 70 + 10 + 2 + 8 ____ + ____ = ____
c. 54 + 37 50 + 30 + 4 + 7 ____ + ____ = ____	**b.** 24 + 55 __+__ + __+__ ____ + ____ = ____
e. 36 + 36 __+__ + __+__ ____ + ____ = ____	**f.** 42 + 68
f. 45 + 18	**h.** 37 + 58

Puzzle Corner   Figure out the missing numbers for these addition problems.

a. ▢▢
+ 1 4
―――
 4 1

b. ▢▢
+   3
―――
 7 1

c. ▢▢
+ 2 5
―――
 5 1

d. ▢▢
+ 7 8
―――
 9 1

e. ▢▢
+ 2 6
―――
 6 1

Adding Three or Four Numbers Mentally

When you add three numbers, you can add them in any order you wish.	Perhaps add 8 and 8 first: $8 + 8 + 6$ $= 16 + 6 = \underline{}$	Or perhaps add 8 and 6 first: $8 + 8 + 6$ $= 8 + 14 = \underline{}$

1. Add three numbers.

a. $8 + 8 + 8 = \underline{}$	b. $7 + 9 + 6 = \underline{}$	c. $5 + 8 + 9 = \underline{}$
d. $7 + 9 + 5 = \underline{}$	e. $8 + 6 + 4 = \underline{}$	f. $2 + 9 + 5 = \underline{}$

When you add four numbers, often it is easy to add them *in pairs:* two numbers at a time.		But sometimes some other way of adding is easier.
Add 7 and 3. Add 5 and 6: $7 + \mathbf{5} + 3 + \mathbf{6}$ $= 10 + 11 = \underline{}$	Add the first two, and the last two: $6 + 9 + \mathbf{8} + \mathbf{5}$ $= 15 + 13 = \underline{}$	Double 8 makes 16, then to that add 4: $9 + \mathbf{8} + \mathbf{8} + \mathbf{4}$ $= 16 + 4 + 9 = \underline{}$

2. Add four numbers. Look at the example.

a. $8 + 8 + 2 + 8$ $= 16 + 10$ $= 26$	b. $7 + 5 + 5 + 6$ $= \underline{} + \underline{}$ $= \underline{}$	c. $4 + 7 + 2 + 5$ $= \underline{} + \underline{}$ $= \underline{}$
d. $6 + 7 + 9 + 8$ $= \underline{} + \underline{}$ $= \underline{}$	e. $8 + 5 + 2 + 6$ $= \underline{} + \underline{}$ $= \underline{}$	f. $4 + 5 + 3 + 9$ $= \underline{} + \underline{}$ $= \underline{}$

3. Practice adding three or four numbers.

a. $4 + 8 + 6 =$ ___	b. $4 + 9 + 5 + 6 =$ ___	c. $7 + 8 + 7 + 9 =$ ___
d. $9 + 9 + 5 =$ ___	e. $8 + 3 + 5 + 4 =$ ___	f. $2 + 6 + 6 + 5 =$ ___
g. $8 + 4 + 4 =$ ___	h. $9 + 2 + 4 + 6 =$ ___	i. $2 + 3 + 8 + 9 =$ ___

4. Madison took photos of her friends. She took eight photos of Mia, nine photos of Chloe, and eight of Isabella. How many photos did Madison take all totaled?

5. Gabriel has 7 toy cars and Lucas has 9. They put their cars together. Can they share the cars evenly? If yes, how many would each boy get?

6. Elijah made 8 sand towers and Bill made 11. Can the boys share the towers in a game they are playing? If yes, how many would each boy get?

7. Add mentally. Think, what would the *easiest order* to add the numbers!

a. $30 + 2 + 40 + 8 =$ ___	c. $9 + 40 + 1 + 4 =$ ___
b. $50 + 4 + 10 + 7 =$ ___	d. $20 + 10 + 8 + 9 =$ ___

8. Compare the expressions and write $<$, $>$, or $=$.

a. $8 + 5 + 6$ ☐ $5 + 6 + 9$	b. $54 + 8$ ☐ $53 + 9$
c. $8 + 8 + 7 + 7$ ☐ $9 + 9 + 6 + 6$	d. $48 - 6$ ☐ $38 + 5$

Adding Three or Four Numbers in Columns

Sometimes we get *two or three new* tens from the ones. We need to regroup.

| In the ones, we add 8 + 7 + 8 = 23.

We write the two new tens in the tens column. Complete the problem. | $$\begin{array}{r} 2 \\ 4\,8 \\ 2\,7 \\ +1\,8 \\ \hline 3 \end{array}$$ | In the ones we add 9 + 9 + 7 + 6 = 18 + 13 = 31. We write *three* new tens in the tens column.

In the tens, we add 3 + 3 + 1 + 2 + 2 = 11. The answer is 111, *more* than one hundred. | $$\begin{array}{r} 3 \\ 3\,9 \\ 1\,9 \\ 2\,7 \\ +2\,6 \\ \hline 1\,1\,1 \end{array}$$ |

1. Add mentally. *Remember* to first try to find if any of the numbers **make 10**.

| a. 8 + 4 + 5 = ___ | b. 3 + 8 + 7 = ___ | c. 8 + 5 + 6 + 4 = ___ |

2. Add. The answers are "hidden" in the list of numbers below the problems.

a. 52
    30
  + 11
  ____

b. 13
    25
  + 54
  ____

c. 33
    38
  + 27
  ____

d. 36
    27
  + 19
  ____

e. 36
    27
    18
  + 16
  ____

f. 40
    18
    16
  + 22
  ____

g. 15
    17
    18
  + 39
  ____

h. 12
    29
    25
  + 14
  ____

i. 19
    69
  + 19
  ____

j. 56
    32
  + 29
  ____

k. 45
    55
  + 19
  ____

l. 59
    19
  + 42
  ____

74   80   82   89   91   92   93   96   97   98   117   107   120   119   122

3. Find the total cost.

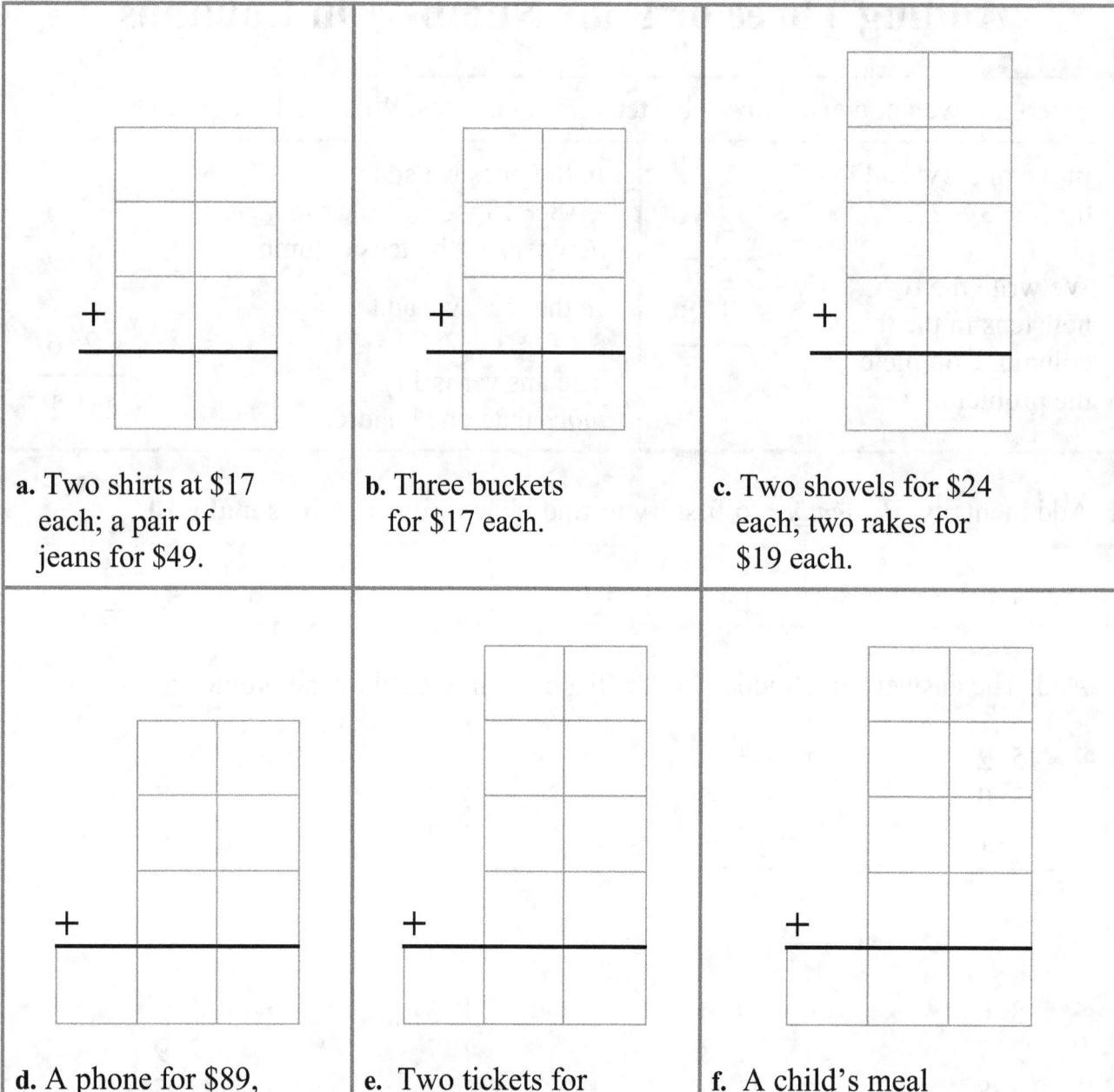

a. Two shirts at $17 each; a pair of jeans for $49.

b. Three buckets for $17 each.

c. Two shovels for $24 each; two rakes for $19 each.

d. A phone for $89, a phone cover for $12, and chocolate for $7.

e. Two tickets for adults for $36 each; two tickets for children for $23 each.

f. A child's meal for $19 and meals for three adults for $29 each.

4. Find the errors in these additions, and correct them.

a.
```
   3 3
 + 4 8
 -----
   711
```

b.
```
   5 5
 + 3 9
 -----
   814
```

5. Solve the problems. You need to add or subtract.

a. One bus has 35 people on it, and another has 22.
How many more people does the first one have than the second?

b. A bus had some people in it. Then, 13 more people got on. Now there are 19 people on the bus.
How many were on the bus originally?

c. One bus can seat 40 people. There were already 33 people.
Is there room for nine more people?

Yes/No, because

d. One bus can seat 40 people.
How many buses do you need for 76 people?

How many buses do you need for 99 people?

e. A bus was full with 40 people, but then six people got off.
How many people are on the bus now?

f. A bus was full with 40 people. First it dropped off 3 people. Then it dropped off seven more people. How many people were left on the bus?

6. Add.

	a.	b.	c.	d.
	3 9	3 3	1 7	5 5
	1 5	4 8	3 7	1 8
	1 8	1 6	2 5	1 5
	+ 2 8	+ 1 3	+ 3 4	+ 2 7

7. Are these numbers even or odd? Mark an "X". If the number is even, write it as a double of some number.

Number	Even?	Odd?	As a double:
8	X		4 + 4
16			
100			
19			

Number	Even?	Odd?	As a double:
18			
24			
15			
21			

Puzzle Corner

Skip-count from 25 (in the middle) to the outer edge. Each sector has a different skip-counting pattern—either by 2s, by 3s, by 4s, by 5s, or by 10s.

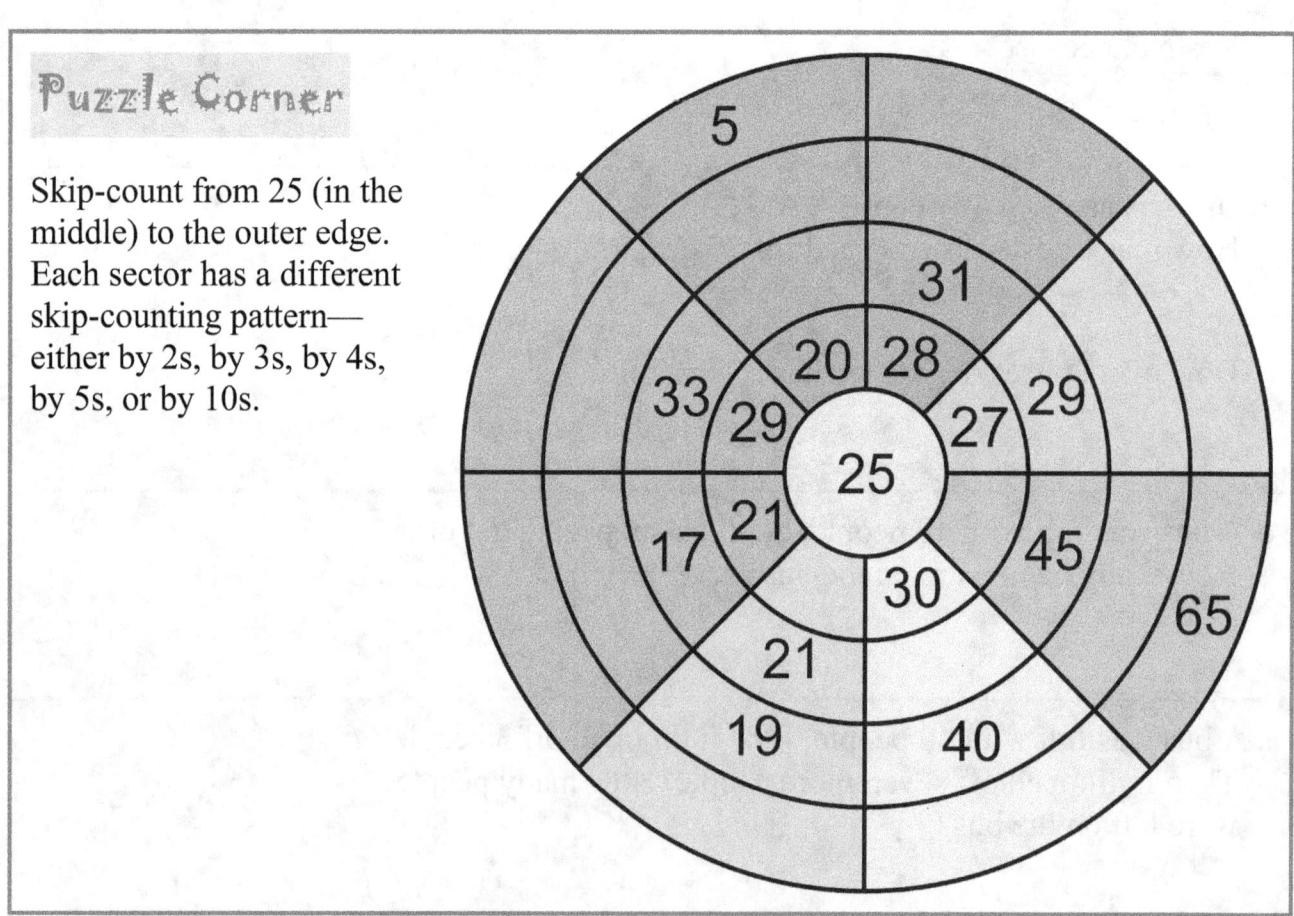

Mixed Review Chapter 4

1. Find one-half and double of the given numbers.

a. $\frac{1}{2}$ of 6 is _____.	b. $\frac{1}{2}$ of 10 is _____.	c. $\frac{1}{2}$ of 8 is _____.
d. Double 6 is _____	e. Double 10 is _____	f. Double 8 is _____

2. Find the number that goes into the shape.

a. 73 + ◯ = 80	b. 78 + ◯ = 98	c. 96 − ◯ = 56
d. ◯ + 92 = 100	e. ◯ − 20 = 5	f. ◯ − 50 = 41

3. Draw a line to connect the problems that are in the same fact family. (You don't need to solve them.)

13 − 8 = ☐	13 − 6 = ☐	15 − 6 = ☐
6 + ☐ = 15	11 − 2 = ☐	13 − 5 = ☐
11 − 9 = ☐	7 + ☐ = 15	5 + ☐ = 11
8 + ☐ = 15	15 − 9 = ☐	15 − 8 = ☐
11 − 5 = ☐	5 + ☐ = 13	9 + ☐ = 11
7 + ☐ = 13	6 + ☐ = 11	☐ + 6 = 13

4. Subtract.

a.	b.	c.	d.
15 − 9 = _____	13 − 9 = _____	14 − 8 = _____	15 − 7 = _____
13 − 6 = _____	14 − 7 = _____	16 − 8 = _____	13 − 5 = _____

5. Write the time.

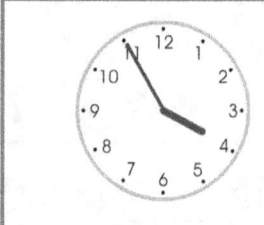			
a. ____ : ____	b. ____ : ____	c. ____ : ____	d. ____ : ____

6. Write the time 10 minutes later than what the clocks show in the previous exercise.

a. ____ : ____	b. ____ : ____	c. ____ : ____	d. ____ : ____

7. Solve the problems.

a. In a game, Kathy got 14 points, Shaun got double that many points, and Abigail got 10 more points than Kathy.
Who got the most points?

How many points was that?

b. You are 8 years old and your brother is double your age.
How many years older is your brother than you?

c. Emma got 7 points more in a game than Matthew.
Matthew got 31 points. How many points did Emma get?

d. One volleyball costs $26 and another costs $6 more than that.
How much does the other volleyball cost?

e. There were 7 more birds in the oak tree than in the birch tree.
If the oak tree had 15 birds, how many birds were in the birch tree?

f. Edith has 12 markers and Judith has 6. They put together their markers and share them evenly. How many does each girl get?

Review Chapter 4

1. Add *in parts*. Break the number that is not whole tens into its tens and ones in your mind.

a. 17 + 10 = _____	b. 16 + 20 = _____	c. 50 + 14 = _____
42 + 10 = _____	67 + 20 = _____	30 + 45 = _____

2. Add.

a. 27 + 8 = _____	b. 18 + 9 = _____	c. 5 + 87 = _____
54 + 7 = _____	73 + 8 = _____	7 + 88 = _____

3. Add by adding tens and ones separately.

a. 36 + 22	b. 72 + 18
30 + 20 + 6 + 2	70 + 10 + 2 + 8
_____ + _____ = _____	_____ + _____ = _____
c. 54 + 37	d. 24 + 55
50 + 30 + 4 + 7	___ + ___ + ___ + ___
_____ + _____ = _____	_____ + _____ = _____

4. Solve the problems.

a. Diane and Ted picked fruit for Mr. Mohan. Diane earned $25 and Ted earned double that. How much did Ted earn?

How much did the two earn together?

b. Emily has 24 flower plants in her yard. Leah has half that many. How many flower plants does Leah have?

5. Add.

a. 43 + 28	b. 33 + 39	c. 24 + 47	d. 23 + 38	e. 55 + 17

f. 38
   13
+ 42
	g. 39
	   10
	+ 46
		h. 41
		   44
		+ 36
			i. 38
			    7
			   49
			+ 23
				j. 27
				   36
				   19
				+ 35

6. Solve.

a. Naomi bought some potatoes for $18, onions for $15, and meat for $40. What was the total cost?	b. If you buy three chairs for $34 each, what is the total bill?	c. Anna has 29 stickers and so does Betty. Ruth has 22 and Judy has 26. How many stickers are there total?

d. Andy had $47 in his wallet. He earned $15 by selling lemonade. Now can he buy a remote-controlled toy car for $65?

If yes, how much money would he have left after buying it?

If not, how much more money would he need?

Chapter 5: Geometry and Fractions
Introduction

This chapter covers geometry topics and an introduction to fractions. In geometry, the emphasis is on exploring shapes. Children learn to identify triangles, rectangles, squares, quadrilaterals, pentagons, hexagons, and cubes, and they draw various basic shapes.

We also study some geometric patterns, have surprises with pentagons and hexagons, and make shapes in a tangram-like game. Overall, the idea is to allow children to explore geometry in fun ways, and help them to learn the terms for basic shapes.

The lesson on rectangles and squares introduces the idea of a shape consisting of many little squares. This idea is continued in the lesson "Some Fractions". The child divides a rectangle into halves or fourths, and counts the little squares in the entire shape and in the fractional part. Naturally, this is building towards the concept of area.

In the lessons on fractions, the child divides basic shapes into halves, thirds, and fourths (quarters). The common notation for fractions is introduced. The last topic of the chapter is comparing fractions using visual models.

Pacing Suggestion for Chapter 5

Please add one day to the pacing for the test if you use it.

The Lessons in Chapter 5	page	span	suggested pacing	your pacing
Shapes Review	130	*3 pages*	1 day	
Surprises with Shapes	133	*2 pages*	1 day	
Rectangles and Squares	135	*3 pages*	1 day	
Making Shapes	138	*2 pages*	1 day	
Geometric Patterns	140	*3 pages*	1 day	
Solids	143	*2 pages*	1 day	
Printable Shapes	145	*4 pages*		
Some Fractions	153	*3 pages*	2 days	
Comparing Fractions	156	*2 pages*	1 day	
Mixed Review Chapter 5	158	*2 pages*	1 day	
Review Chapter 5	160	*2 pages*	1 day	
Chapter 5 Test (optional)				
TOTALS		*28 pages*	11 days	

Games and Activities

Pattern Blocks/Tangram

All children love tangram games, or using shapes to make new composite shapes. The list of further Internet resources gives links to free online versions. The below links give examples of sets on Amazon, but there are many more available. Look for a set that comes with pattern cards.

Coogam Wooden Pattern Blocks Set 130Pcs
https://www.amazon.com/dp/B07MYYK64R/?tag=mathmammoth-20

LOVESTOWN Wooden Pattern Blocks Set 230 Pcs
https://www.amazon.com/dp/B08C37JMP2/?tag=mathmammoth-20

Connect the Dots

You need: paper, pencil, ruler

Draw four to eight dots randomly on the paper, and number them. Then draw another set of dots and label those with a, b, c, etc. Ask the child to connect the numbered dots in order, to get a closed shape, and also, separately, the dots with letters. Two shapes will be formed that may overlap. See an example in the lesson "Surprises with Shapes", questions 10-12.

- After the child has the basic idea, ask the child to name the shapes *before* drawing them.
- Try to draw dots in such a manner that you get two overlapping rectangles, or other patterns.
- Take turns, so that the child also draws and labels the dots, and you draw the shapes.

Color Patterns

You need: graph paper (a grid of squares) and triangular paper

Ask the child to play around with colorful patterns using squares and rectangles in the square grid paper, and using triangles, hexagons, and other shapes in the triangular paper. The child should repeat the pattern over the page or part of the page. Give examples. The patterns don't have to consist of same-size shapes, but they may.

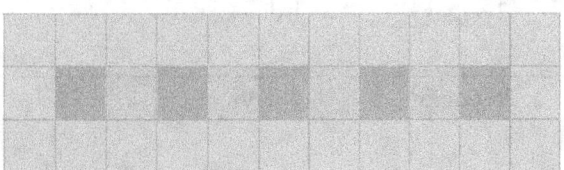

Download graph paper with squares:
https://www.mathmammoth.com/download/square-grid-paper.pdf

Download triangular graph paper:
https://www.mathmammoth.com/download/triangular-grid-paper.pdf

Rectangles with Squares

You need: graph paper. You can download a page of it below:
https://www.mathmammoth.com/download/square-grid-paper.pdf

Ask the child to draw a rectangle with a given number of little squares inside it. Use numbers less than 20. Occasionally, use a small prime number, such as 5, 7, or 11, so the child can find out that there is only one way to draw such a rectangle (5 by 1, 7 by 1, 11 by 1 rectangle).

Take turns, so that the child also asks you to draw a rectangle with a given number of squares. This activity prepares the student for the concepts of area and multiplication.

Rectangle Fractions

You need: graph paper. You can download a page of it below:
https://www.mathmammoth.com/download/square-grid-paper.pdf

This is a continuation of the previous activity (Rectangles with Squares). Ask the child to draw a rectangle with a given number of little squares inside it, and then divide it into halves or quarters. Use numbers less than 20.

Questions to ponder:

- What happens if you choose an odd number?
- What happens when you divide a rectangle with 6, 10, or 14 squares into quarters?

Games and Activities at Math Mammoth Practice Zone

Fraction Matcher
Match visual models of fractions to the written forms. Choose "Fractions" and "Level 1".
https://www.mathmammoth.com/practice/fraction-matcher

Further Resources on the Internet

These resources match the topics in this chapter, and offer online practice, online games (occasionally, printable games), and interactive illustrations of math concepts. We heartily recommend you take a look. Many people love using these resources to supplement the bookwork, to illustrate a concept better, and for some fun. Enjoy!

https://l.mathmammoth.com/gr2ch5

Shapes Review

1. Draw three dots on the right. Connect the dots with straight lines. You have drawn a **triangle** (*tri* means *three*).

   It has _____ vertices (corners) and three sides.

   Draw two more triangles in the same space. They can overlap.

2. Draw FOUR dots on the right. Connect the dots with straight lines. You have drawn a **quadrilateral** (*quadri* means *four*; *lateral* has to do with sides).

   It has _____ vertices (corners) and four sides.

   Draw two more quadrilaterals in the same space.

3. The figures on the right are a square and a rectangle. Can you tell which is which?

   Squares and rectangles are **quadrilaterals** because they have four sides.

   Draw at least one more square and one more rectangle into the picture, the best you can.

4. Draw FIVE dots on the right.
   Connect the dots with straight lines.

   You have drawn a **pentagon**
   (*penta* means *five*).

   It has _____ vertices and _____ sides.

   Draw one more pentagon in the space.

5. Draw SIX dots on the right. Connect the dots with straight lines.

   You have drawn a **hexagon**
   (*hex* means *six*).

   It has _____ vertices and _____ sides.

   Draw yet one more hexagon in the space.

6. How is *a circle* different from all of the shapes above?

7. Continue the pattern, and color it with pretty colors!

8. Color all triangles yellow.
   Color all quadrilaterals green.
   Color all pentagons blue.
   Color all hexagons purple.

   Or choose your own colors for each kind of shape.

9. Now, this is a challenge to check if you remember the words for different shapes. Don't look at the previous pages! You can use the "dot" method: first draw dots for the corners, then use a ruler to draw the lines connecting the dots.

   a. Draw here a big and a small four-sided shape. What are four-sided shapes called?

   b. Draw here a skinny and a fat three-sided shape. What are three-sided shapes called?

   c. Draw here a blue five-sided shape and a green six-sided shape. What are five and six-sided shapes called?

Surprises with Shapes

1. Connect the dots using a ruler. Be neat! What shape do you get?

   _____

2. Draw a line from one corner to some other corner. This divides your shape into <u>two</u> new shapes. What shapes are they?

   _____

3. Draw more lines from a corner to some corner so that the whole shape gets divided into triangles.

4. Connect the dots using a ruler. Be neat! What shape do you get?

   _____

5. Draw a line from one corner to the opposite corner. Then repeat so that each corner gets connected to the opposite corner. You need to draw three lines to do that.

6. Decorate your shape now so that it becomes a SNOWFLAKE! ALL snowflakes have this basic shape.

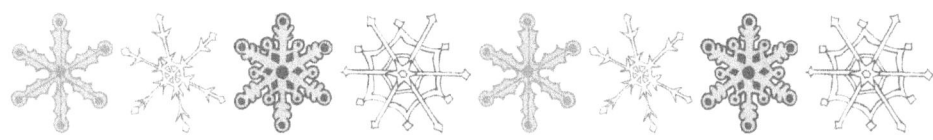

7. Connect the dots *in the numbered order* using straight lines. Be neat! What do you get?

   _____

8. In the middle of that shape, another shape is formed. What is it?

   _____

9. Also connect the dots in the order 1 - 4 - 2 - 5 - 3 - 1. What shape is formed now?

   _____

10. Connect the dots 1-2-3 using a ruler. Then connect the dots a-b-c also. Be neat! What shape do you get?

    _____

11. In the middle of that shape, another shape is formed. What is it?

    _____

12. Also connect the dots in the order 1 - a - 2 - b - 3 - c - 1. What shape is formed now?

    _____

Rectangles and Squares

1. Continue these patterns that use rectangles and squares.

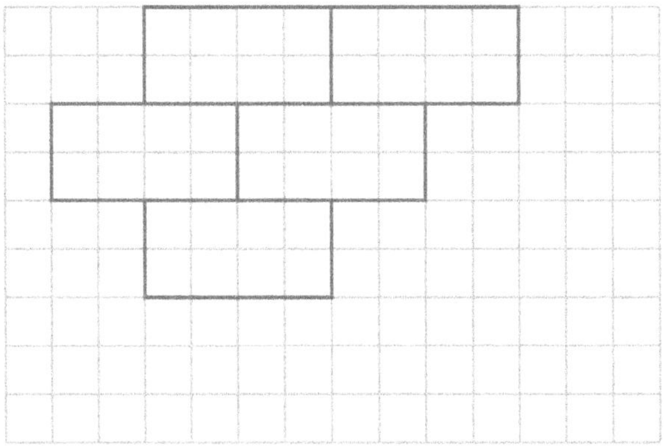

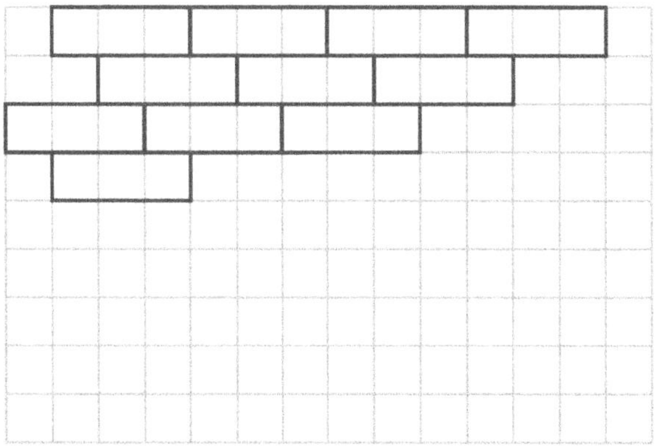

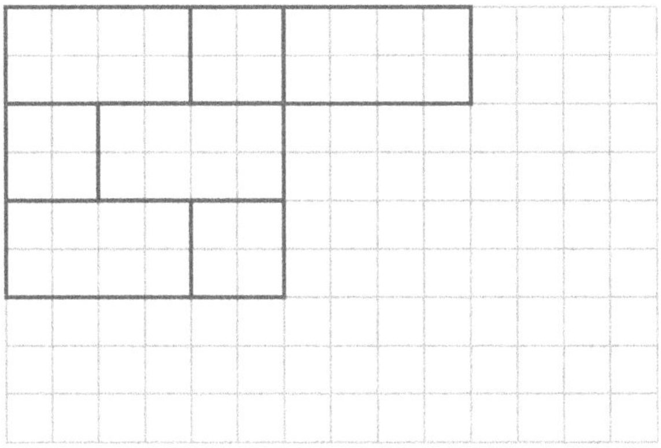

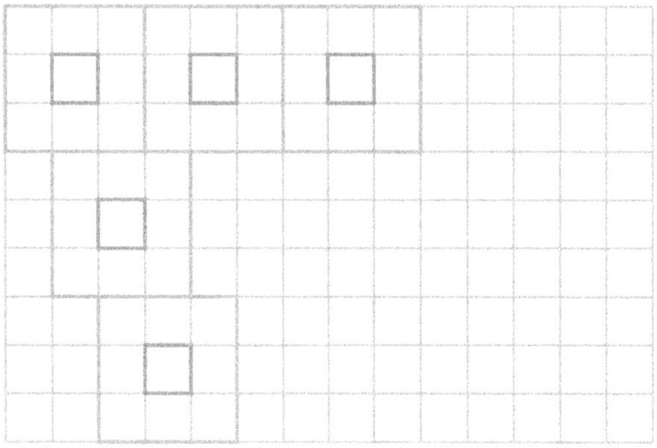

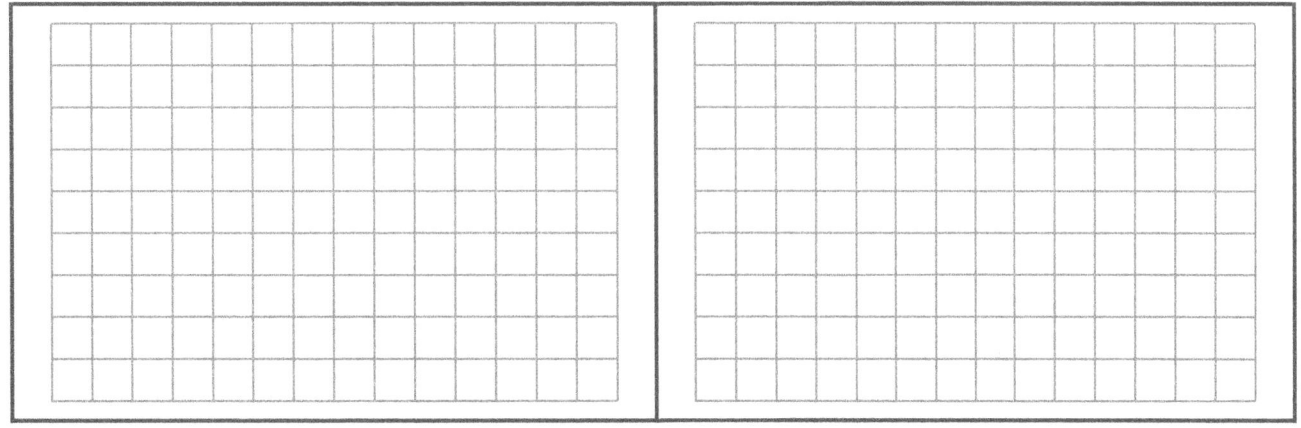

Make your own patterns here!

Josh counted how many little squares were inside this rectangle.
He got 12 little squares.

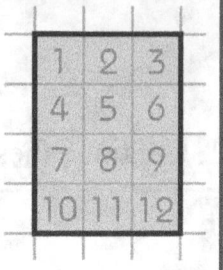

2. Now you do the same. Count how many little squares are inside each rectangle.

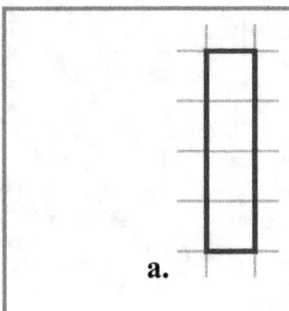

a.

_____ little squares

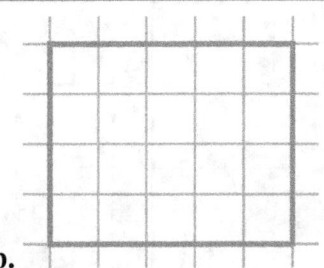

b.

_____ little squares

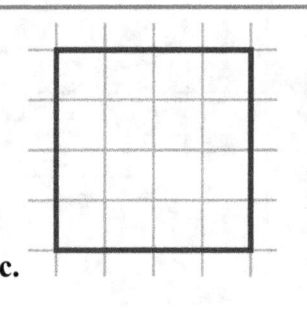

c.

_____ little squares

3. Draw rectangles so they have a certain number of little squares inside. Guess and check!

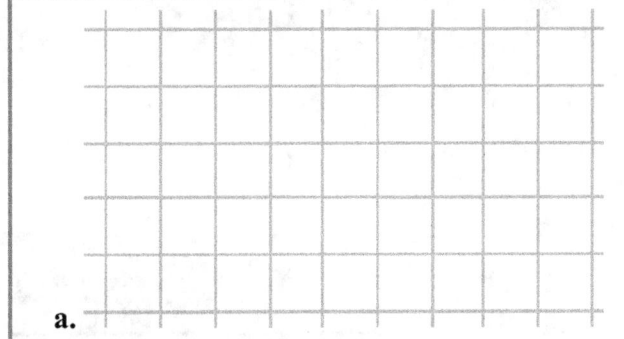

a.

10 little squares

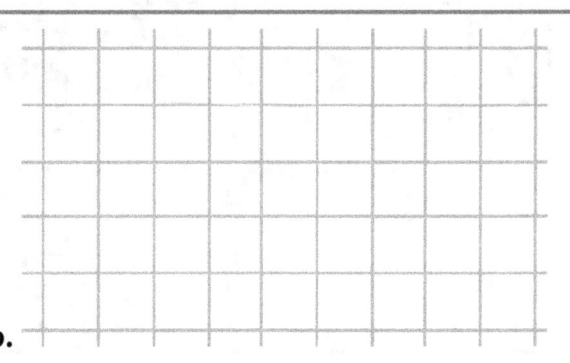

b.

15 little squares

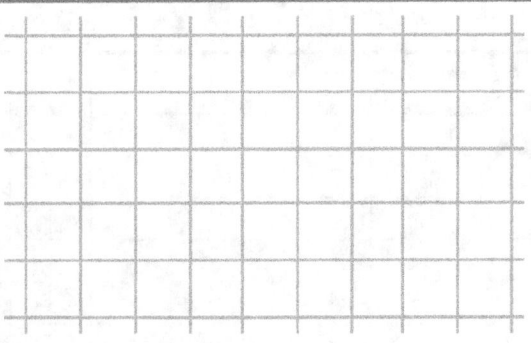

c.

8 little squares
Can you make two different ones?

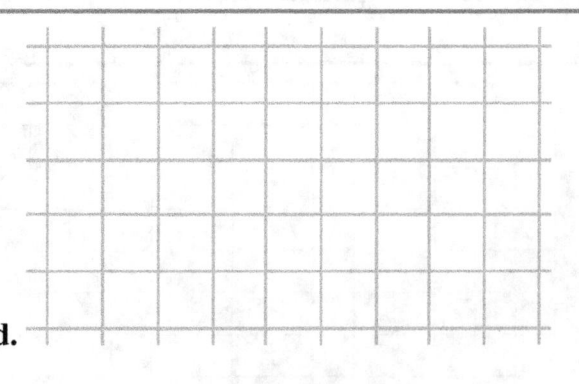

d.

12 little squares
Can you make two different ones?

4. Here is a pattern where several squares are inside each other. Continue the pattern. Use pretty colors.

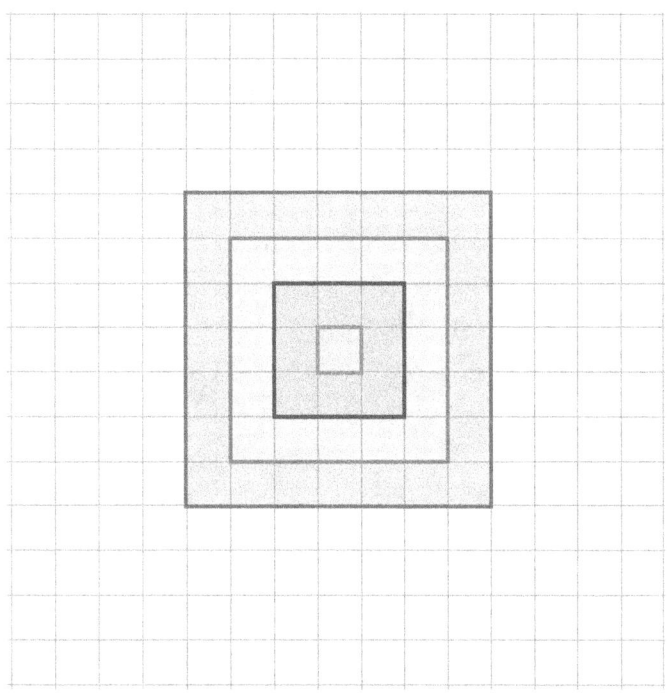

5. Design your own pattern, where you start with a small rectangle in the middle, then draw bigger ones around it like in the pattern above.

Making Shapes

We can make new shapes from putting several shapes together. For example, these two triangles together form a square:

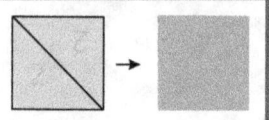

1. Cut out the shapes on the next page. What shapes can you use to make the given shapes? There may be several possible solutions. The figures below are smaller than the ones you will cut out.

   a.   b.

   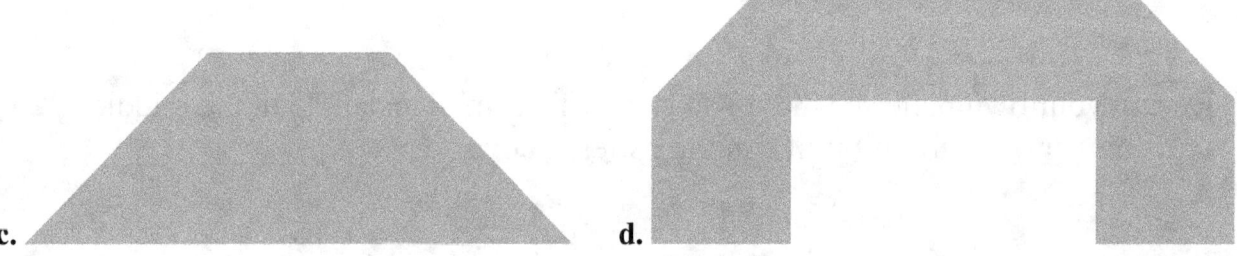

   c.   d.

   e.   f.

2. Now, you do the same. Put together some shapes. Trace the outline of your combined shape on paper, and give that to your friend to solve.

3. The game you just played is very similar to the ancient Chinese puzzle called Tangram. Play an interactive tangram game online at
   https://www.mathplayground.com/tangrams.html or

   https://www.abcya.com/tangrams.htm

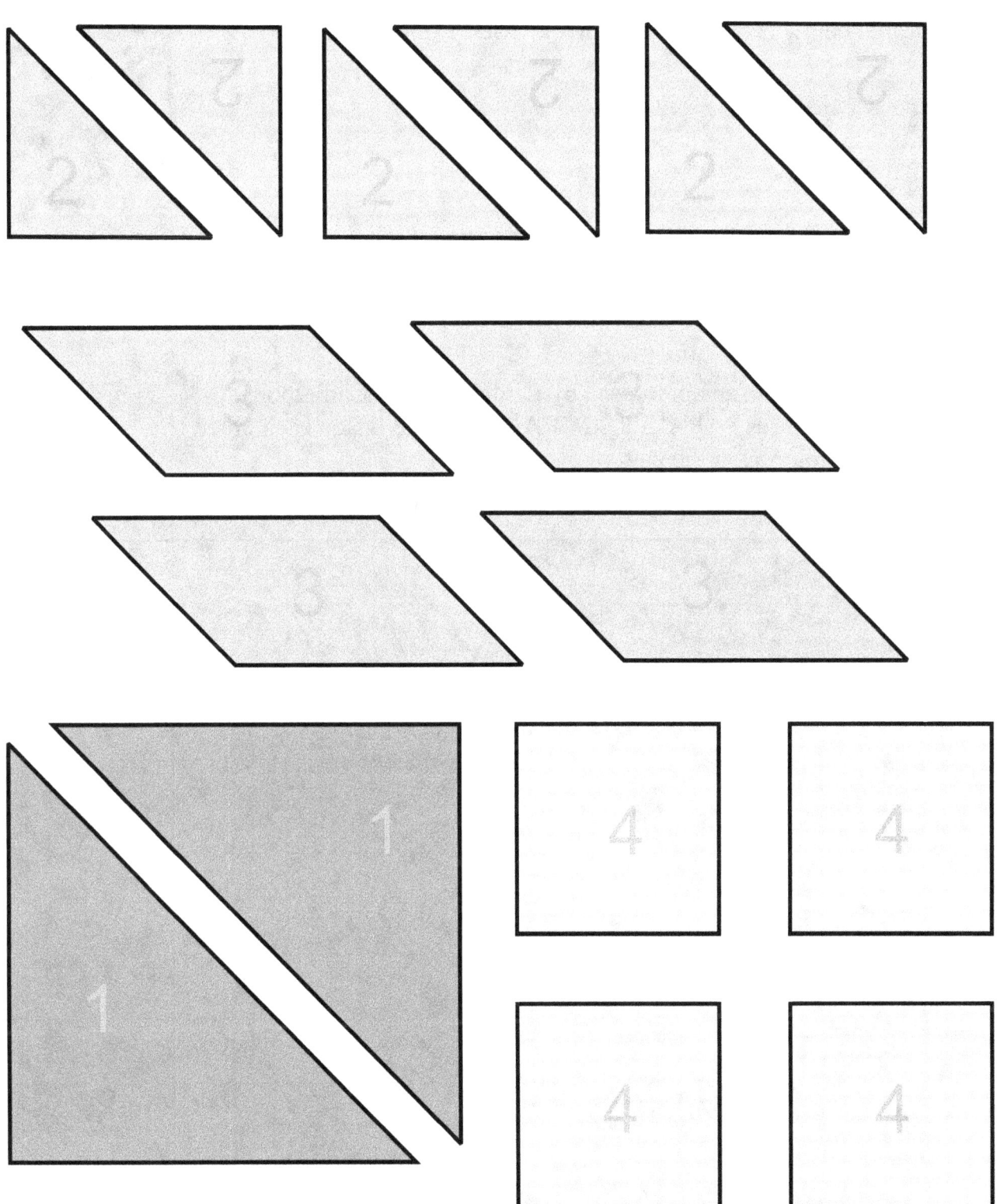

Geometric Patterns

1. The design below is often seen in Greek vases. Continue it.

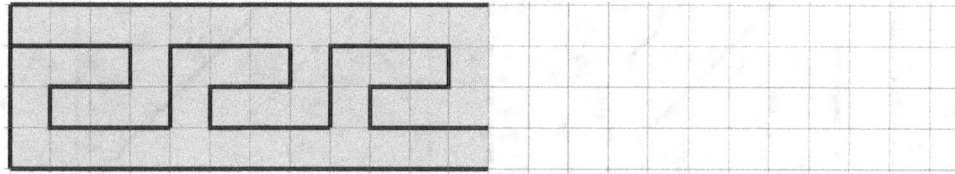

2. This is a pattern from an apron used by Kirdi people in Cameroon, Africa. Notice it uses PARALLELOGRAMS that are inside each other. Continue the coloring in the pattern.

140

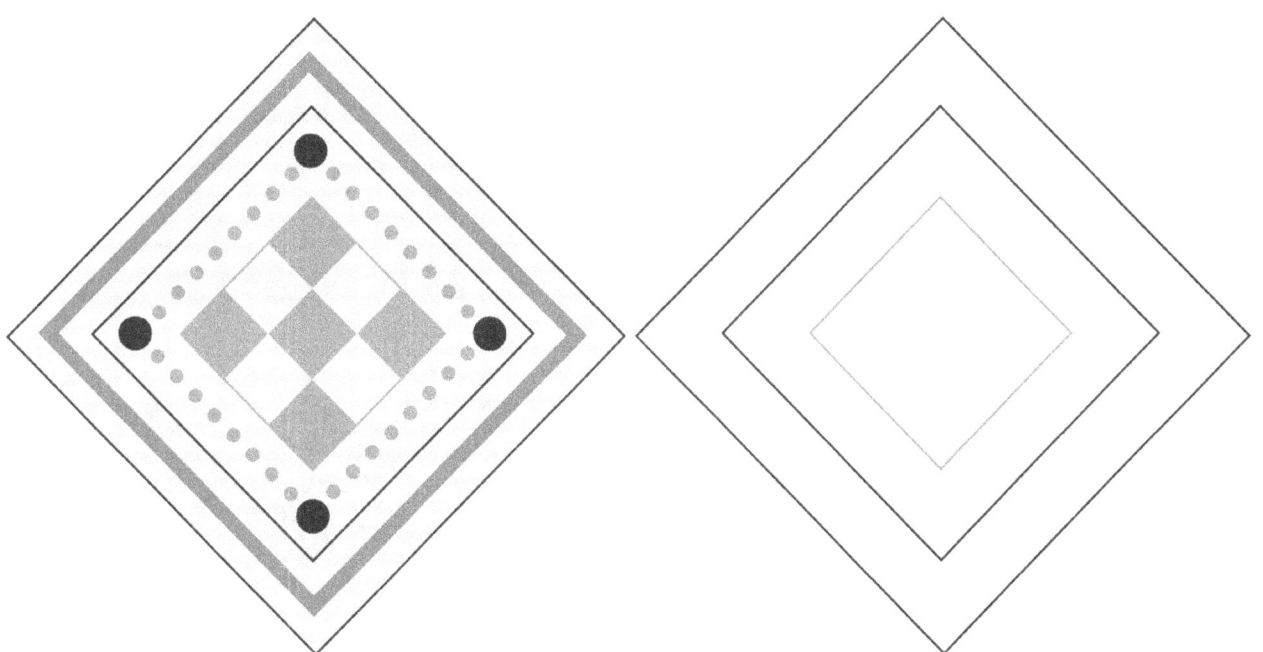

3. This is a geometric design found on a Greek vase.

   a. What two shapes are used in this design?

   _____ and _____

   b. Copy the design at least once in the empty shapes.

4. Repeat the patterns to fill the grids.

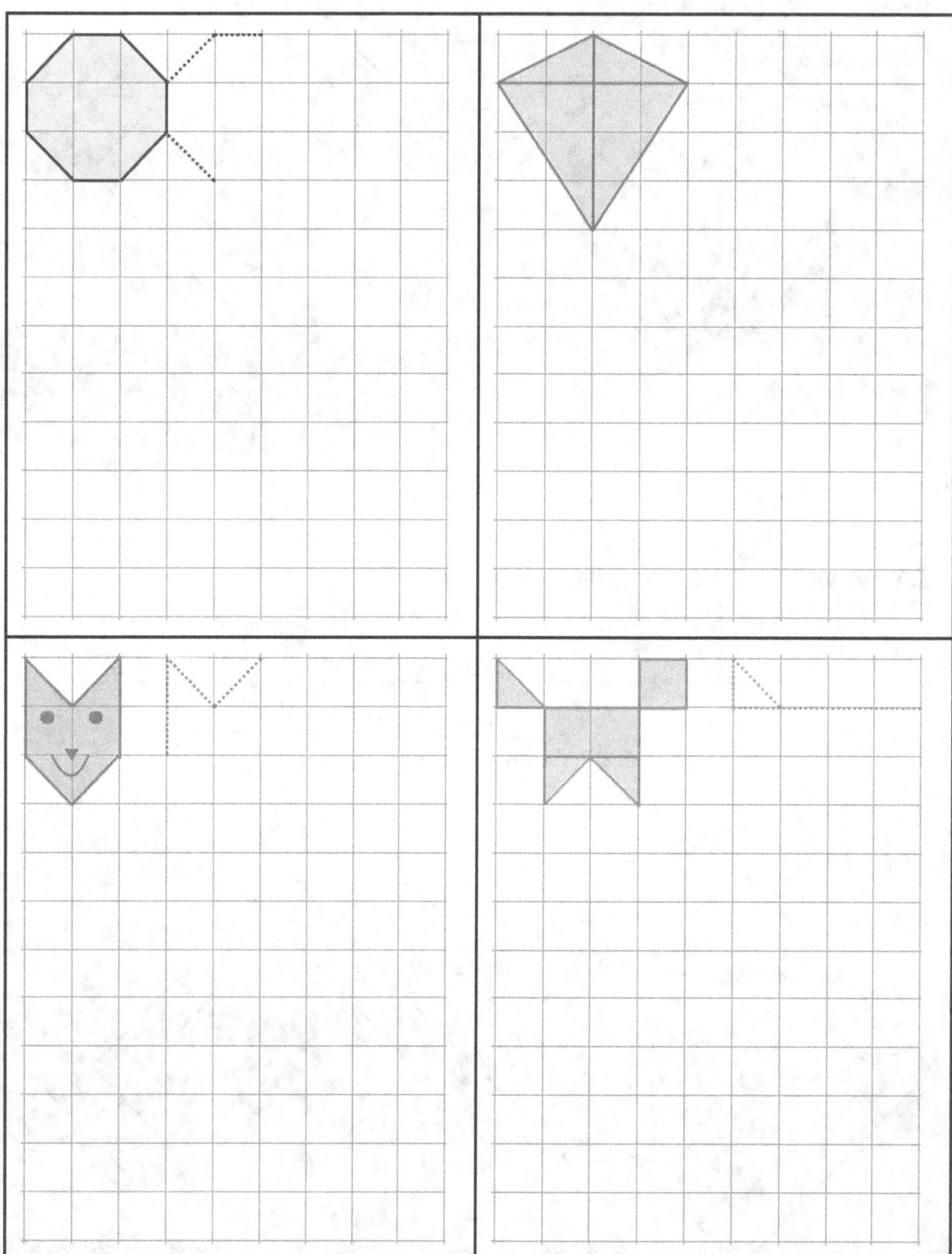

142

Solids

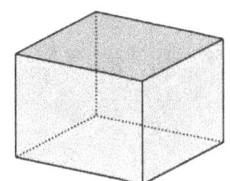 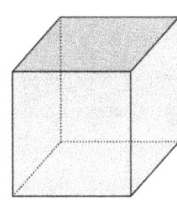 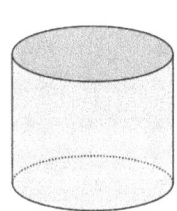 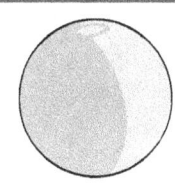

This is a **box**. It is also called a "rectangular prism."

A **cube** is a box, too, but all of its sides are equal in length.

A **cylinder** has a circle on the bottom and at the top.

This is a **sphere**, or just a ball.

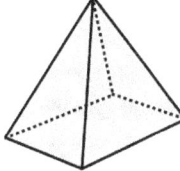 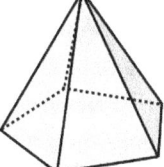  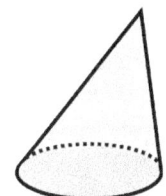 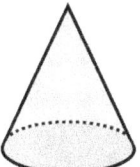 

A pyramid has a pointed top. Its bottom shape can be any many-sided figure, such as a triangle, a rectangle, a square, or a pentagon.

A cone has a pointed top, as well, but it has a rounded shape on the bottom.

1. Make a cube, a cylinder, a cone, and a pyramid using the cut-outs provided on the following pages. Your teacher will help you.

2. A *face* is any of the flat sides of a solid.

   a. Count how many faces a cube has. _____ faces

   What shapes are they?

   b. Count how many faces a box has. _____ faces

   What shapes are they?

   c. Count how many faces this pyramid has.

   _____ faces

   What shapes are they?

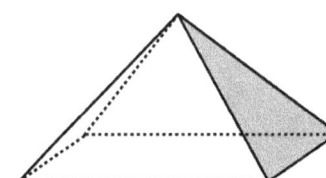

3. You might have seen *safety cones* on the street. They are used to mark off areas where people are not supposed to go. Can you think of other things in real life that are in the shape of a cone, or a part of them is a cone?

_____

_____

(Hint: One thing that is cone-shaped tastes really yummy!)
(Hint: Another thing you might see in birthday parties.)

4. Label the pictures with *box*, *cube*, *cylinder*, *pyramid*, or *cone*.

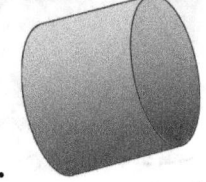  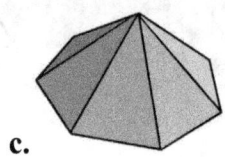

a. _____ b. _____ c. _____

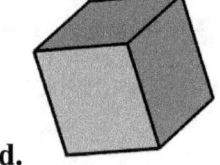 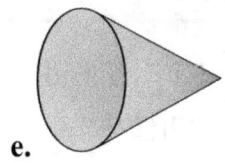 

d. _____ e. _____ f. _____

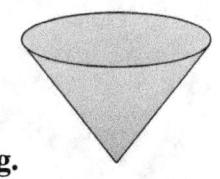 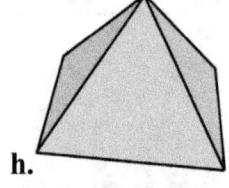 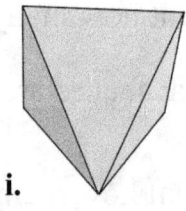

g. _____ h. _____ i. _____

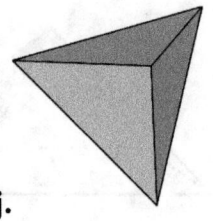  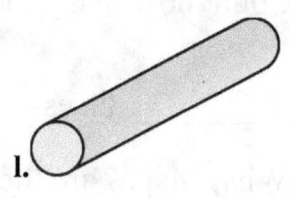

j. _____ k. _____ l. _____

Cube Cut-out

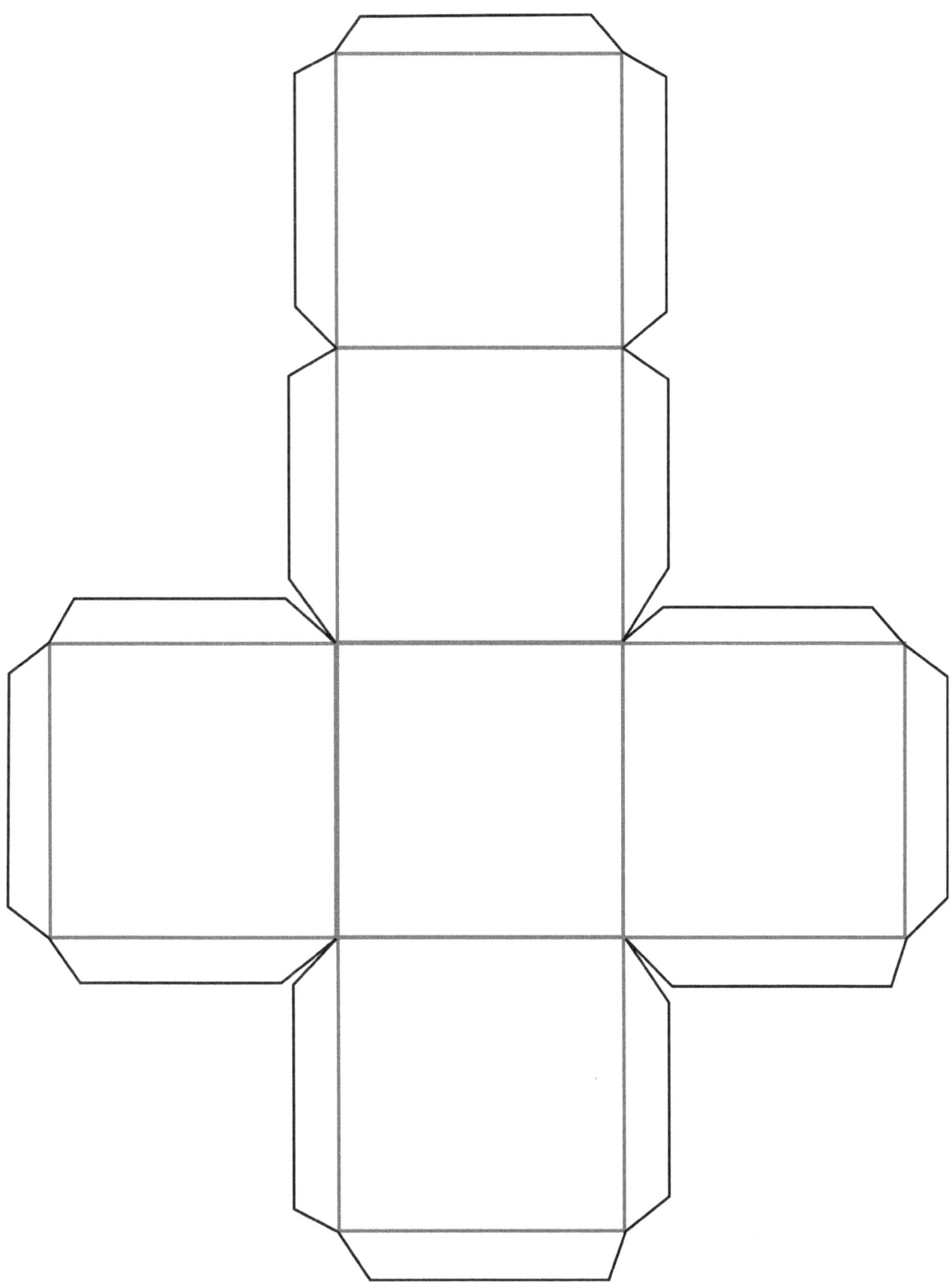

[This page is intentionally left blank.]

Cone Cut-out

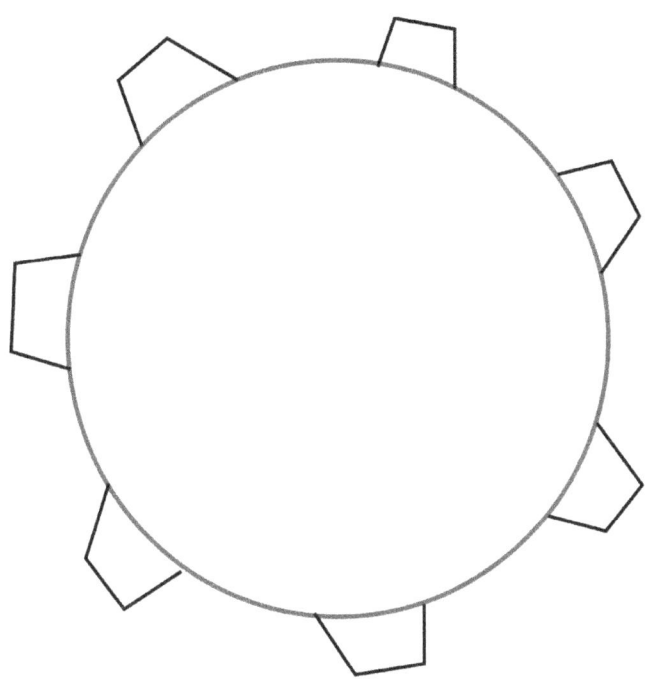

A party hat is an excellent example of a cone. With this cutout, it is easy to tape or glue together the main body of the cone, but attaching the circle to the end is not easy.

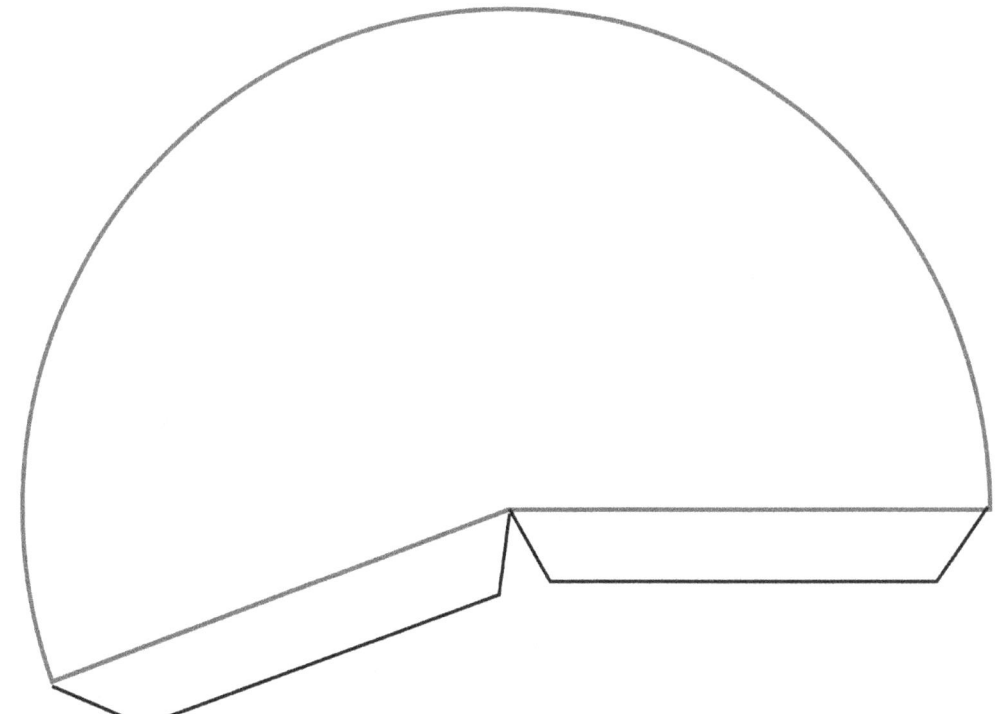

[This page is intentionally left blank.]

Cylinder Cut-out

It might be easier to use a toilet paper roll as a model for a cylinder than to cut and glue/tape this cut-out together. However, putting this together will help the student to understand that the "body" of the cylinder is in the shape of a rectangle.

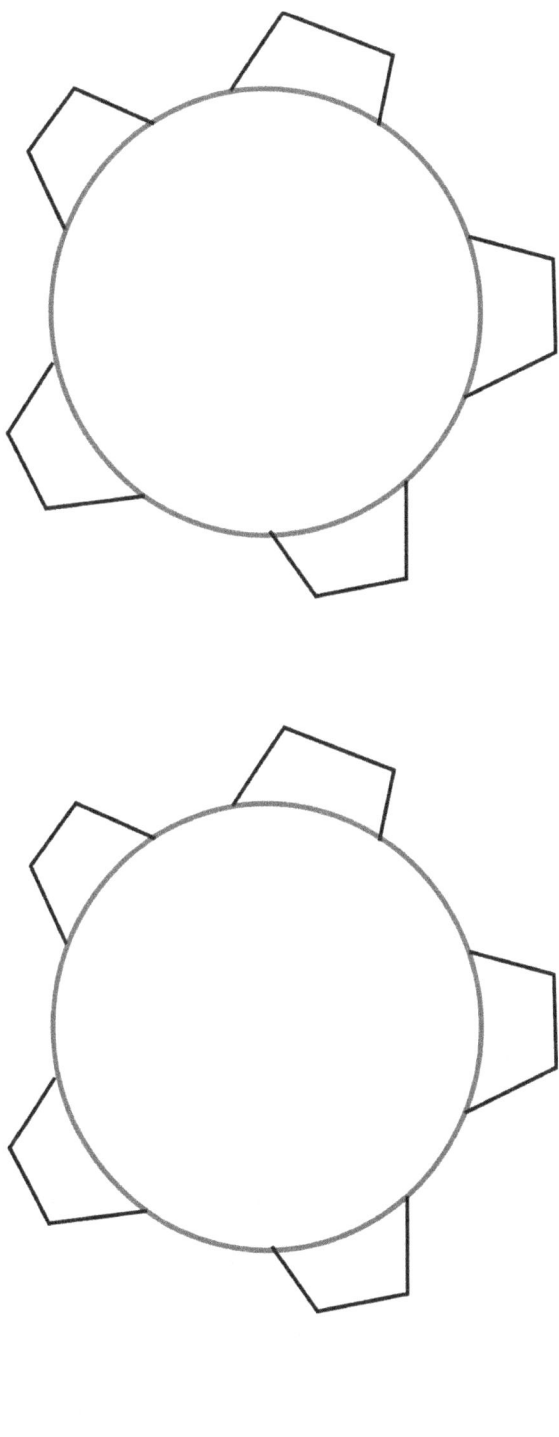

[This page is intentionally left blank.]

Square Pyramid Cut-out

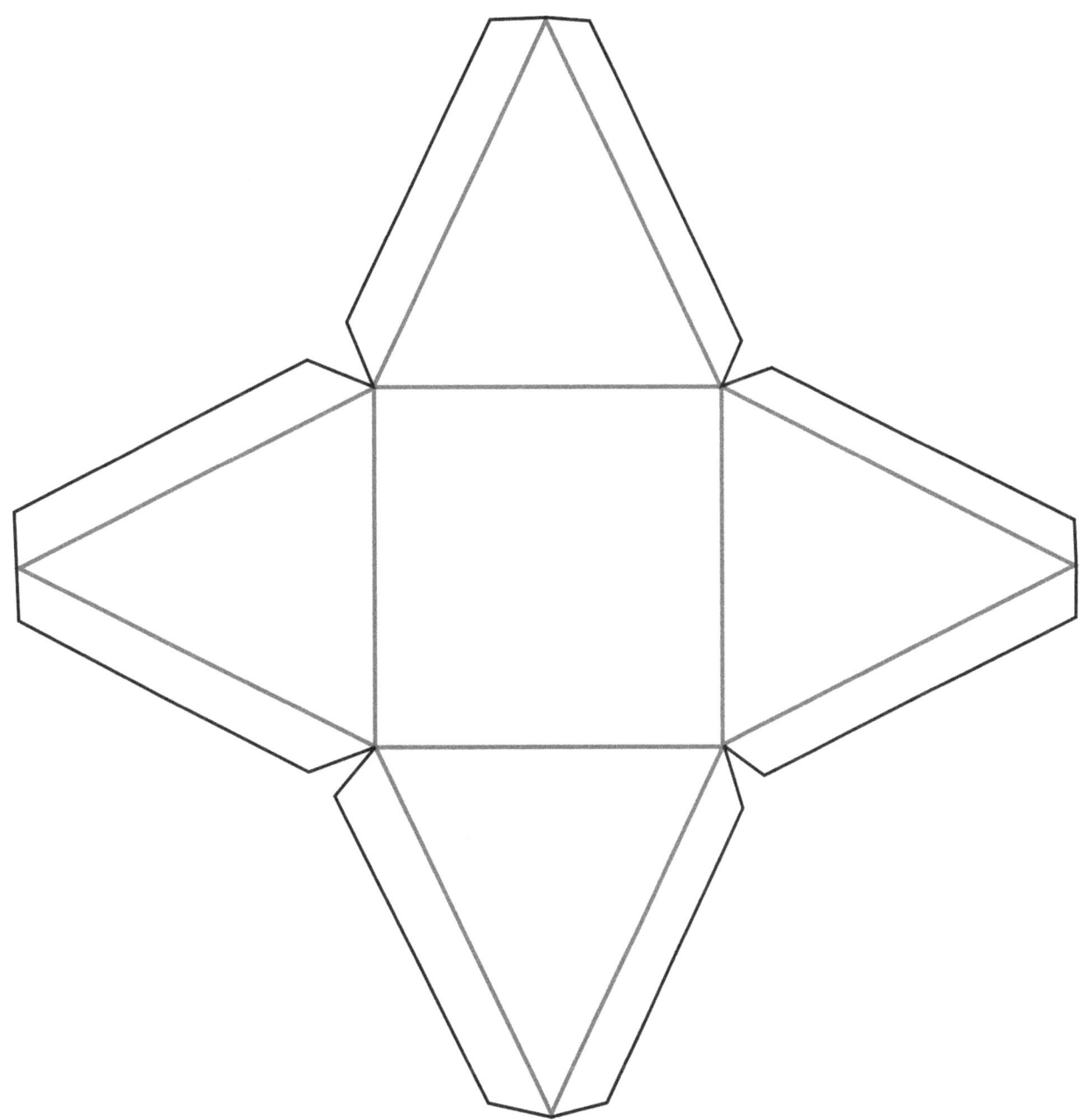

[This page is intentionally left blank.]

Some Fractions

We will now divide shapes into EQUAL parts = parts that are the same size.
When we divide something into TWO equal parts, the parts are called **halves**.
When we divide something into THREE equal parts, the parts are called **thirds**.
When we divide something into FOUR equal parts, the parts are **fourths** or **quarters**.

Here, <u>one-half</u> of the square is colored. We write $\frac{1}{2}$ or 1/2.	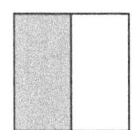	Here, <u>two halves</u> of the square are colored. We write $\frac{2}{2}$ or 2/2. This is the same as 1 (one whole).	
Here, <u>one-third</u> of the square is colored. We write $\frac{1}{3}$ or 1/3.	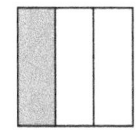	Now, <u>four quarters</u> of the circle are colored. We write $\frac{4}{4}$ or 4/4. This is the same as 1 (one whole).	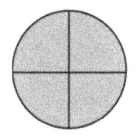

In a **fraction,** we use two numbers, one on the top and one on the bottom.

One-fourth of the pie is colored. how many parts colored → $\frac{1}{4}$ how many equal parts →	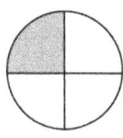	Two-thirds of the square is colored. how many parts colored → $\frac{2}{3}$ how many equal parts →	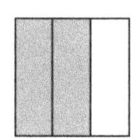

1. Divide these shapes. Then color as you are asked to.

a. 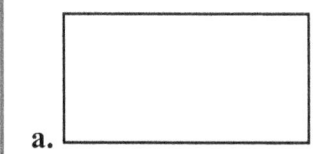 Divide this into halves. Color $\frac{1}{2}$	b. 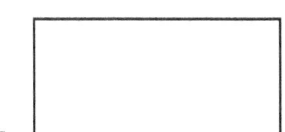 Divide this into thirds. Color $\frac{1}{3}$	c. 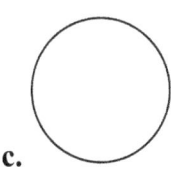 Divide this into halves. Color $\frac{2}{2}$	d. 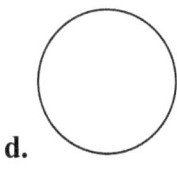 Divide this into fourths. Color $\frac{2}{4}$
e.  Divide this into quarters. Color $\frac{4}{4}$	f. 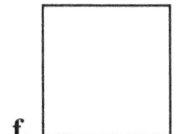 Divide this into thirds. Color $\frac{2}{3}$	g. 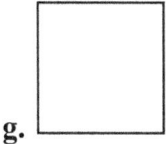 Divide this into fourths. Color $\frac{1}{4}$	h. 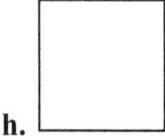 Divide this into halves. Color $\frac{2}{2}$

Robert divided this square into fourths, and then colored $\frac{1}{4}$ of it.

<u>Notice</u>: the whole rectangle has 16 **little squares** inside it.

The fourth that Robert colored has 4 **little squares** inside it.

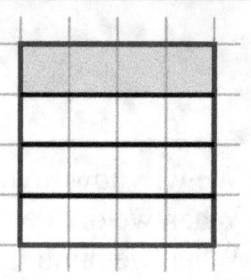

2. Complete.

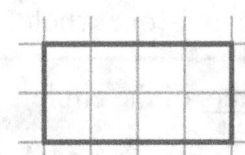

a.

Divide this into halves. Color $\frac{1}{2}$

_____ little squares in one half.

_____ little squares in the whole rectangle

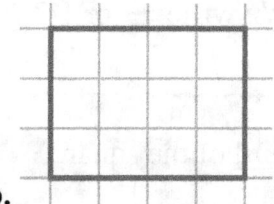

b.

Divide this into halves. Color $\frac{1}{2}$

_____ little squares in one half.

_____ little squares in the whole rectangle

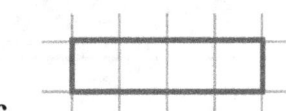

c.

Divide this into fourths. Color $\frac{1}{4}$

_____ little squares in one fourth.

_____ little squares in the whole rectangle

d.

Divide this into fourths. Color $\frac{1}{4}$

_____ little squares in one fourth.

_____ little squares in the whole rectangle

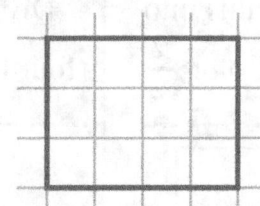

e.

Divide this into fourths. Color $\frac{3}{4}$

_____ little squares in three fourths.

_____ little squares in the whole rectangle

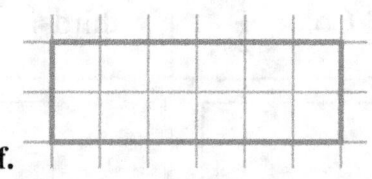

f.

Divide this into thirds. Color $\frac{2}{3}$

_____ little squares in two thirds.

_____ little squares in the whole rectangle

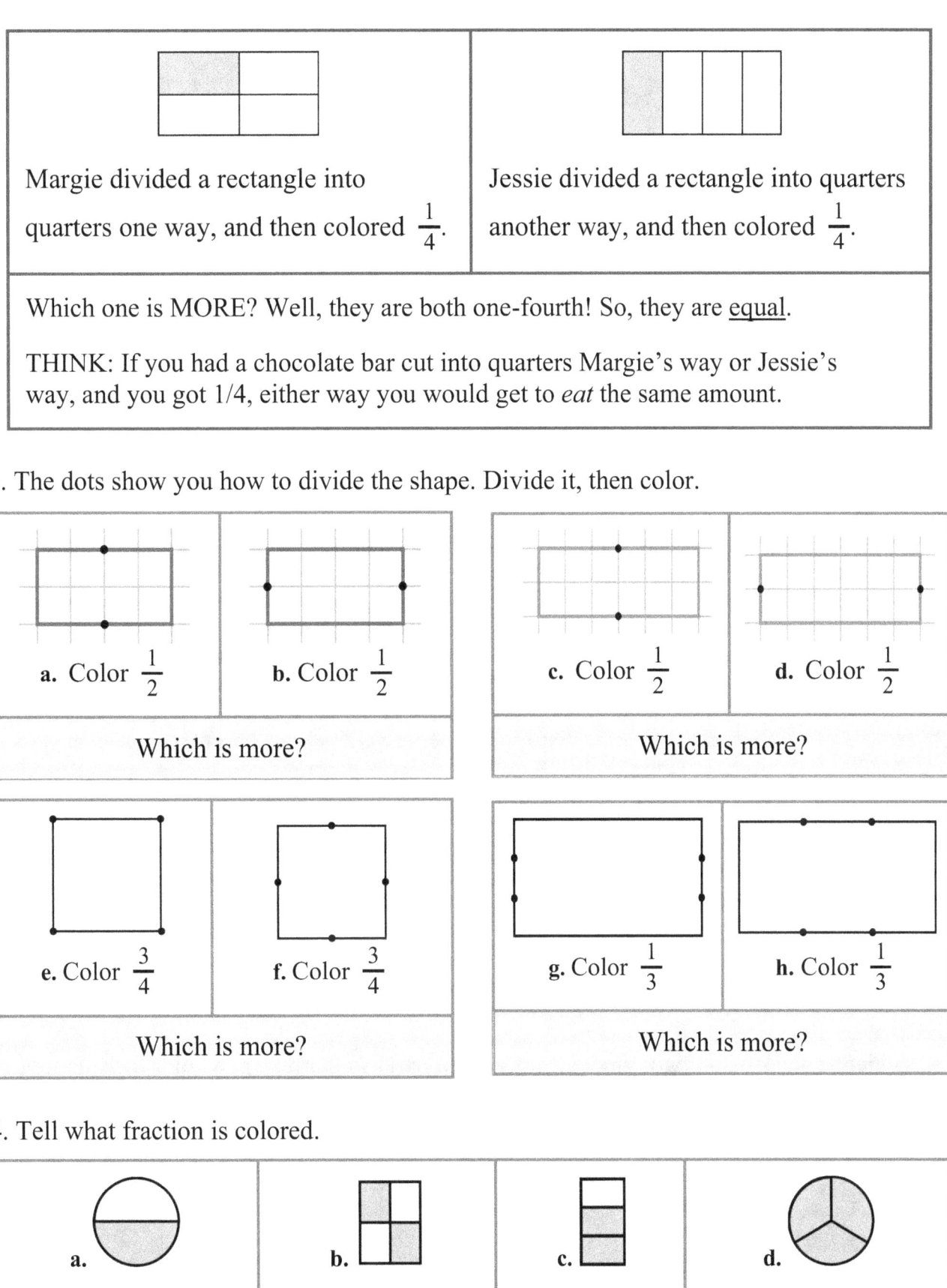

Comparing Fractions

1. Color the whole shape. Then write 1 whole as a fraction. Lastly, read what you wrote with numbers.

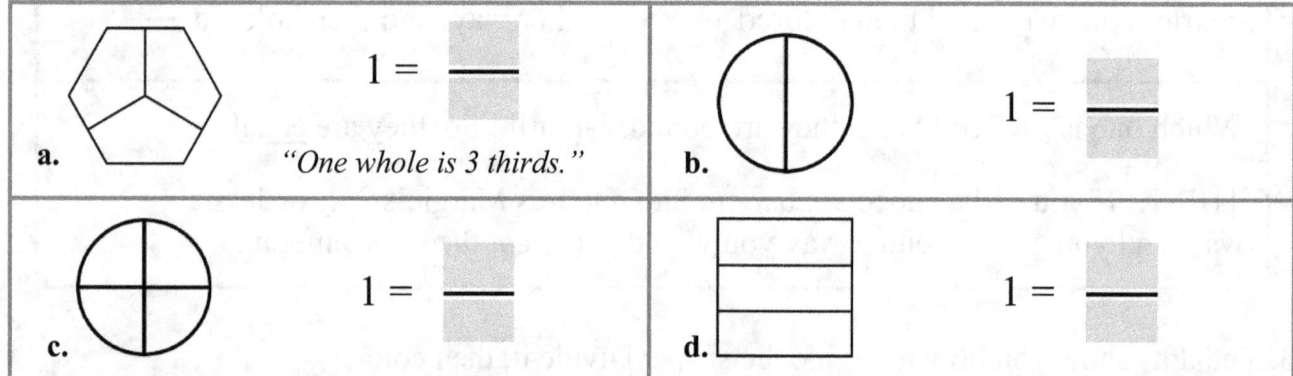

a. $1 = \dfrac{}{}$ "One whole is 3 thirds."

b. $1 = \dfrac{}{}$

c. $1 = \dfrac{}{}$

d. $1 = \dfrac{}{}$

2. Color. Then compare and write < , > , or = . Which is more "pie" to eat?

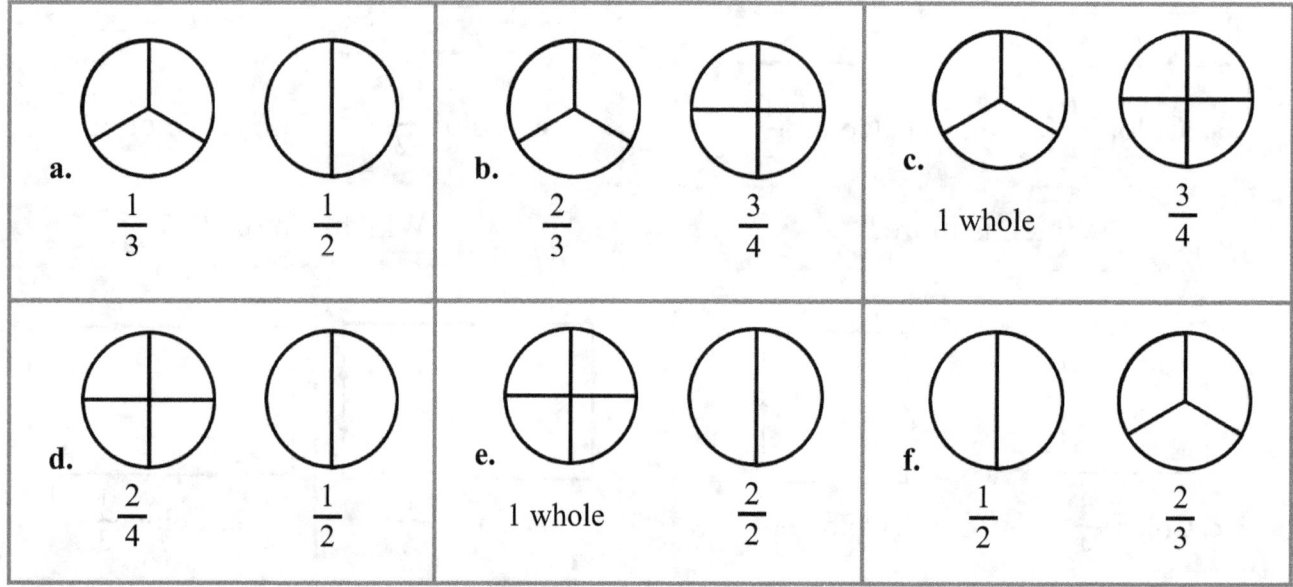

a. $\dfrac{1}{3}$ $\dfrac{1}{2}$

b. $\dfrac{2}{3}$ $\dfrac{3}{4}$

c. 1 whole $\dfrac{3}{4}$

d. $\dfrac{2}{4}$ $\dfrac{1}{2}$

e. 1 whole $\dfrac{2}{2}$

f. $\dfrac{1}{2}$ $\dfrac{2}{3}$

3. Divide the shapes into two, three, or four equal parts so that you can color the fraction. Then compare and write < , > , or = .

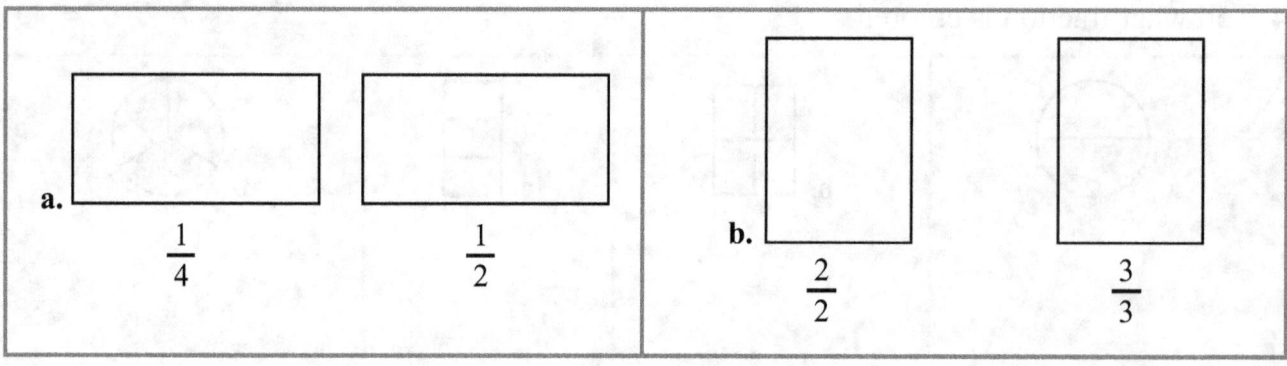

a. $\dfrac{1}{4}$ $\dfrac{1}{2}$

b. $\dfrac{2}{2}$ $\dfrac{3}{3}$

More fractions

When we divide something into FIVE equal parts, the parts are called *fifths*.
When we divide something into SIX equal parts, the parts are called *sixths*.

Here, <u>five-sixths</u> of the square is colored. We write $\frac{5}{6}$ or 5/6. 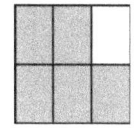	Here, <u>two fifths</u> of the circle are colored. We write $\frac{2}{5}$ or 2/5. 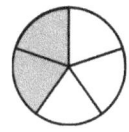

4. Color the given fraction.

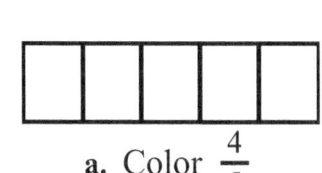 a. Color $\frac{4}{5}$

 b. Color $\frac{2}{5}$

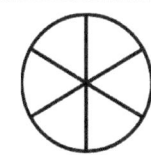 c. Color $\frac{5}{6}$

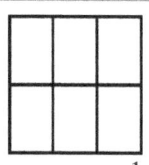 d. Color $\frac{1}{6}$

5. Color. Then compare and write <, >, or =. Which is more "pie" to eat?

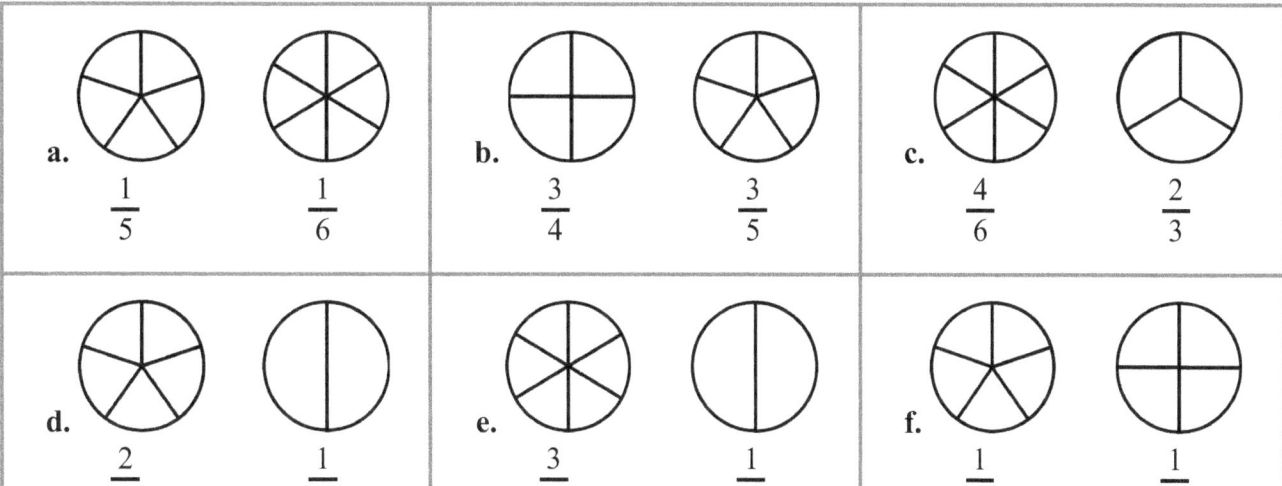

6. Divide the shapes into two, three, or four equal parts so that you can color the fraction. Then compare and write <, >, or =.

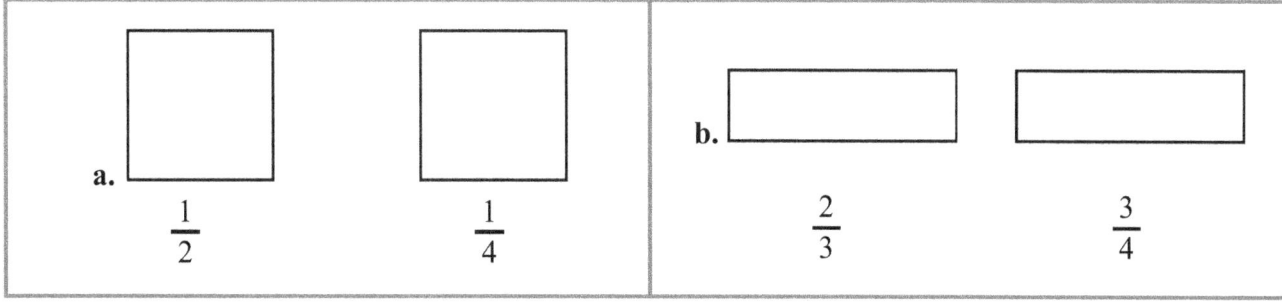

Mixed Review Chapter 5

1. Find the differences.

a. The difference of 100 and 95 _____	b. The difference of 40 and 20 _____
c. The difference of 16 and 8 _____	d. The difference of 56 and 4 _____

2. Subtract. Think of the difference.

a. 25 − 22 = _____	b. 76 − 71 = _____	c. 51 − 49 = _____

3. Find the missing numbers.

a. 14 − ☐ = 5	b. 13 − ☐ = 8	c. 16 − ☐ = 9
d. ☐ − 6 = 6	e. ☐ − 7 = 4	f. ☐ − 4 = 9

4. Add. Compare the problems.

a. 8 + 3 = _____ 18 + 3 = _____	b. 6 + 6 = _____ 86 + 6 = _____	c. 8 + 7 = _____ 48 + 7 = _____
d. 46 + 7 = _____	e. 47 + 9 = _____	f. 88 + 5 = _____

5. Add. Regroup the ones to make a new ten.

```
a.            b.           c.           d.           e.
    6 4          4 7          1 3          1 5          2 7
    1 5          2 7          5 6          2 6            9
  + 2 5        + 2 3        + 2 6          4 7          3 5
  _____        _____        _____        + 1 9        + 2 5
                                          _____        _____
```

6. Find how much the things cost together.

a. a fishing rod, $38 baits, $9 bucket, $8	b. skis, $79 jacket, $22 socks, $11	c. four chairs, $29 each
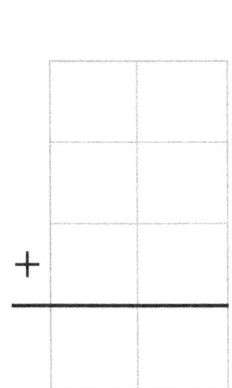	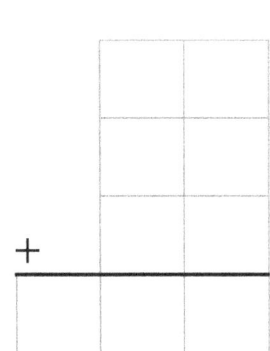	

7. Add four numbers. Look at the example.

a. $8 + 8 + 2 + 8$ $= 16 + 10$ $= 26$	b. $9 + 5 + 5 + 8$ = ____ + ____ = ____	c. $6 + 7 + 3 + 5$ = ____ + ____ = ____
d. $7 + 7 + 8 + 8$ = ____	e. $9 + 4 + 4 + 7$ = ____	f. $6 + 4 + 4 + 9$ = ____

8. Solve the problems. You need to add or subtract.

a. One bike costs $78, and another costs $23 more than the first. Find the price of the second bike.	b. One shirt costs $29 and another costs $15. How much more does the first shirt cost than another?	c. You bought both shirts in problem (b). How much did they cost together?
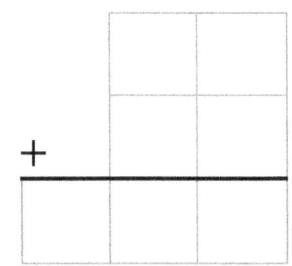	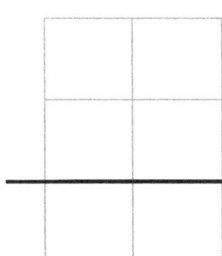	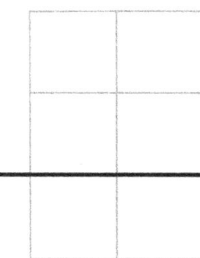

Review Chapter 5

1. Connect the dots. Use a ruler!
   What shape do you get?

   _____

2. Choose one corner of your shape.
   Now draw a line (with a ruler)
   from that corner to some other
   corner so that you will divide the
   shape into a <u>triangle</u> and a <u>pentagon</u>.

3. Draw in the grid a square that
   has 4 little squares inside.

4. Draw in the grid a rectangle that
   has 18 little squares inside.

5. What is this shape called? _____

   How many <u>faces</u> does it have? _____

   What shape are the faces? _____

6. Sarah put together these two triangles. What new shape did she get?

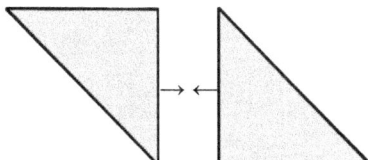

7. Label the pictures as *box*, *cylinder*, *pyramid*, or *cone*.

   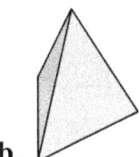   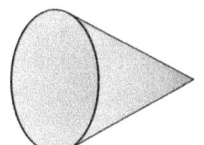

a. _____   b. _____   c. _____

8. Color the whole shape. Then write 1 whole as a fraction. Lastly, read what you wrote with numbers.

a. 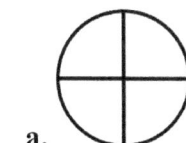   1 = ——

b.    1 = ——

9. Divide the shapes into two, three, or four equal parts so that you can color the fraction.

a. 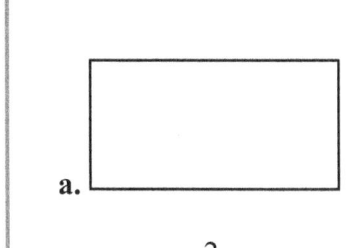   b. 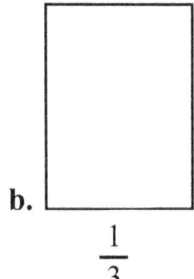   c.    d.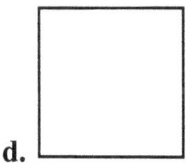

$\frac{2}{4}$ $\frac{1}{3}$ $\frac{2}{3}$ $\frac{2}{2}$

10. Color. Then compare and write < , > or = . Which is more "pie" to eat?

a. 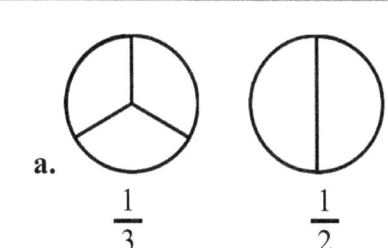   b. 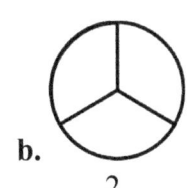    c.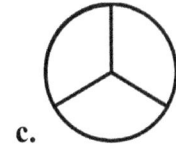

$\frac{1}{3}$ $\frac{1}{2}$ $\frac{2}{3}$ $\frac{3}{4}$ 1 whole $\frac{3}{4}$

www.ingramcontent.com/pod-product-compliance
Lightning Source LLC
Chambersburg PA
CBHW081237180526
45171CB00005B/451